T0243100

Crossing the Bridge of the Digital Divide: A Walk with Global Leaders

A Volume in:
Educational Leadership for Social Justice

Series Editors

Jeffrey S. Brooks
Denise E. Armstrong
Ira Bogotch
Sandra Harris
Whitney Sherman Newcomb
George Theoharis

Educational Leadership for Social Justice

Series Editors

Jeffrey S. Brooks
RMIT University

Denise E. Armstrong
Brock University

Ira Bogotch
Florida Atlantic University

Sandra Harris
Lamar University

Whitney Sherman Newcomb
Virginia Commonwealth University

George Theoharis
Syracuse University

Crossing the Bridge of the Digital Divide: A Walk with Global Leaders

Edited by

Anthony H. Normore
Antonia Issa Lahera

INFORMATION AGE PUBLISHING, INC.
Charlotte, NC • www.infoagepub.com

Library of Congress Cataloging-In-Publication Data

The CIP data for this book can be found on the Library of Congress website (loc.gov).

Paperback: 978-1-64113-390-6
Hardcover: 978-1-64113-391-3
eBook: 978-1-64113-392-0

CONTENTS

FOREWORD

Bridging Divides in Cyber-Lives

Jabari Mahiri

Co-editors Anthony H. Normore and Antonia Issa Lahera have gathered a diverse, international group of thought-leaders to illuminate opportunities and challenges for bridging the digital divide. The multidimensional approaches of these scholars are essential because considerations for achieving digital equity are not only myriad, but mercurial. I discuss digital "divides" in the plural rather than singular to signal the complex and highly changeable ways they are manifested in people's lives. Statistics from the Pew Research Center's "Mobile fact sheet" (2017) indicate that 95% of Americans now own some kind of cell phone (77% of which are smartphones). Globally, according to "Internet usage and world population statistics" (2017), more than half of the world's population now has some level of Internet usage. Increasingly, throughout the world we are leading cyber-lives. But these seemingly hopeful statistics conceal complicated techno-social, techno-economic, and techno-political motives and forces that can make digital equity a bridge too far. Importantly, this book's *Walk with Global Leaders* clarifies both the possibilities and problems of *Crossing the Bridge of the Digital Divide.*

Roots of the digital age are in the shifts from mechanical and analogue to digital technologies that began in the 1950s. However, most scholars agree that

Crossing the Bridge of the Digital Divide: A Walk with Global Leaders, pages ix–xii.

the digital age really took off in the 1970s with the proliferation of computers. A feature of these new digital tools was that they could translate multiple forms of information and media into numerical data. Turning diverse "texts" like pictures, graphics, moving images, and sounds, as well as writing into computable data allowed them to be converged, configured, and communicated in novel ways. These "new media" are characterized by their greatly increased accessibility, mobility, and interchangeability and also by how they have magnified and often simplified processes of meaning production and propagation. The emergence of new media has enabled and necessitated new literacies and new identities. The integrative, remixing capabilities of digital tools allow for creation and experimentation with multisensory, interactive, cyber-experiences that virtually transform the sense of self (Mahiri, 2011, 2017). Crucially, the emergence of rapidly changing forms of new media has also resulted in new relationships between those who primarily produce it and those who mainly consume it.

Consequently, digital divides are reflected in unequal access to, creation and use of, and impacts from the new media and meanings generated and disseminated by information and communication technologies (ICT). But there are additional intricacies of digital divides within and between countries and global regions. Within countries divides exist among individuals, families, communities, businesses, institutions (like schools or service agencies), and geographic areas (like rural and urban). Between countries global divides exist with regard to the state and availability of technology and infrastructure, Internet access and coverage, levels of education and digital literacy, and barriers or affordances of national languages on-line. Major inequalities exist, for example, in the distribution of installed telecommunication bandwidth with only 10 countries (primarily the U.S., China, and Japan) containing around 75% of the global telecommunication capacity as of 2014 (Hilbert, 2016). When the Federal Communications Commission voted in December of 2017 to dismantle "net neutrality" regulations instituted by the Obama administration in 2015, possibilities for equal access were definitively diminished because broadband providers would no longer be prohibited from blocking websites or charging more for higher quality/higher speed service or content. Unless efforts underway to repeal this decision are successful, techno-economic motives of Internet providers will surely decrease the equality of online access and experiences.

One way to synthesize and compare key aspects of digital divides is to look at *who* connects *how* to *what*. When looking at *who* connects, characteristics of individuals, organizations, countries, or global regions intersect with considerations of age, income, education, religion, ethnicity, and gender/sexual orientation among other factors. When looking at *how* connections are made, considerations of motivation, duration, and location also come into play. Are connections made for information, communication, or entertainment? Are they made for social, gaming, political, or work activities? When looking at *what* the specific digital devices are that are being used to connect, considerations of how smart the devices are, whether they are fixed or mobile, and whether they are inter-connected or separate in their use are important. So *who* connects *how* to *what* in conjunction with the amount of bandwidth that's both available and used yields a wide-range of issues regarding digital divides. Yet, these issues are more complex and nuanced.

For example, although the U.S., China, and Japan together had more than 50% of the installed bandwidth potential in 2014 (Hilbert, 2016), based on a report from the International Telecommunications Union (2013) these countries were not among the top 15 in the percentage of their populations using the Internet in 2012. This report showed an exponential rise in Internet usage for people in many countries from 2000 to 2012, but the top 15 countries all had above 85% Internet usage and included places like Iceland (96%), Norway (95%), and Sweden (94%). Interestingly, the number one country was the Falkland Islands (97%). Others that might not be expected to be above the U.S. at 81% were Bermuda (91%), Bahrain (88%), and Qatar (88%). China had the highest number of people using the internet in 2012, however, that number reflected only 42% of its population. On the lower extreme, countries like Ethiopia, Eritrea, Somalia, Niger, and Guinea had less than 2% of their populations using the Internet. These disparities are critical for understanding the global nature of the digital divide. Additionally, one's language can significantly increase or limit access and use of the Internet, despite digital translation capabilities. The top ten languages spoken by web users in order from highest to lowest are English, Chinese, Spanish, Arabic, Portuguese, Japanese, Russian, Malay, French, and German—a tiny fraction of the world's 7,100 living languages.

In addition to the considerations already noted, digital divides cannot be bridged without a significant increase in digital literacies internally within countries and across the regions of the world. Particularly, in this post-truth era of techno-political propaganda and techno-social fake news, the roles of educators, researchers, and policy-makers are paramount in defining and guiding the intellectual, ethical, and socio-emotional development of cyber-citizens in an increasingly digital and duplicitous world. What is the nature of digital justice locally and globally? Why is digital capital as unevenly distributed as economic and social capital? Can we protect ourselves and our privacy when watching and using digital devices that are also watching and using us? Will the digital revolution result in greater freedom and inclusion, or greater oppression and marginalization? In crossing the bridge of the digital divide, how can global leaders insure that cultural traffic travels in both directions? These and other urgent questions and issues are cogently addressed in this important and timely book.

—Dr. Jabari Mahiri
Professor of Education, University of California Berkeley
William and Mary Jane Brinton Family Chair in Urban Teaching

REFERENCES

Hilbert, M. (2016). The bad news is that the digital access divide is here to stay: Domestically installed bandwidths among 172 countries for 1986–2014. *Telecommunications Policy, 40,* 6.

Internet usage and world population statistics. (June 30, 2017). Miniwatts Marketing Group. Retrieved January 27, 2018. http://www.internetworldstats.com/stats.htm

Mahiri, J. (2011). *Digital tools in urban schools. Mediating a remix of learning.* Ann Arbor, MI: University of Michigan Press.

Mahiri, J. (2017). *Deconstructing race: Multicultural education beyond the color bind.* New York, NY: Teachers College Press.

Mobile fact sheet. (January 12, 2017). *Internet and technology report.* Washington, DC: Pew Research Center. Retrieved January 29, 2018. http://www.pewinternet.org/fact-sheet/mobile/

International Telecommunications Union. (2013). *Percentage of individuals using the internet 2000–2012.* Geneva, Switzerland: Author. Retrieved January 25, 2018. https://www.itu.int/en/ITU-D/Statistics/.../2013/Individuals_Internet_2000–2012.xls

INTRODUCTION

Anthony H. Normore and Antonia Issa Lahera

The combined effect of the rapid growth of information is an increasingly fragmented information base, a large component of which is available only to people with money and/or acceptable institutional affiliations. In the recent past, the outcome of these challenges has been characterized as the "digital divide" between the information "haves" and "have nots" along racial and socio economic lines that seem to widen as time passes (del Val & Normore, 2008). To address the issues of digital equity and digital inequality in an effort to bridge the digital divide, educational scholars and practitioners are in positions to ensure equitable opportunities are made available for people of all ages, races, ability, sexual orientation, and ethnicity in support of social justice for bridging the digital divide. The digital divide addresses issues concerning equal opportunity, equity and access that have an effect on the development of marginalized and otherwise disenfranchised populations within and across systems (Pleasant & Ritzhaupt, 2014) nationally and internationally.

THE DIGITAL DIVIDE

The digital divide addresses issues concerning equal opportunity, equity and access that have an effect on the development of marginalized and otherwise disenfranchised students in education systems (Carvin, 2006; Pleasant & Ritzhaupt, 2014). The term entered the collective psyche in the 1990s with national and

Crossing the Bridge of the Digital Divide: A Walk with Global Leaders, pages xiii–xix.
Copyright © 2019 by Information Age Publishing
All rights of reproduction in any form reserved.

global implications. It describes the inability to access computer technology and the use of the internet by certain segments of society (i.e. the poorest) from participation in the global information society (Picciano, 2006).

This socioeconomic digital divide is a global phenomenon that encompasses a widening equity gap in primary, secondary, and higher education throughout the world (Carvin, 2006; Hohlfeld, Ritzhaupt, & Barron, 2013). The divide affects one-sixth of the world's population—approximately 1 billion people (del Val & Normore, 2008). Low-income families face added disadvantages because of lack of access to computers in the home (inability to pay for equipment and services). Consequently, in terms of social justice and information technology *leadership* has become more critical than ever as discourses about the knowledge economy focus on the necessity of educating ALL students with skills for the global workplace. Clearly however, populations with limited access become less prepared for the increasingly global market that continue to emerge in the 21st century (Hohlfeld, Ritzhaupt, & Barron, 2013; Murrell, 2006; Picciano, 2006).

SOCIAL JUSTICE LEADERSHIP

To address the issues of the digital divide, school leaders are in a position to ensure opportunities are made available for all students in schools. Important conceptual research (e.g., Bogotch, 2014) suggests that a social justice orientation to educational leadership practice and research can address how institutionalized theories, norms, and practices in schools and society lead to social, economic, technological, and educational inequities. According to Marshall and Oliva, (2006), researchers and practitioners have embraced the moral imperative of improving "practice and student outcomes for minority, economically disadvantaged, female, gay/lesbian, and other students who have not traditionally served well in schools" (p. 6). As a result, the influence of leadership activity on institutional racism, gender discrimination, inequality of opportunity, inequity of educational processes, digital exclusion, and justice have gained currency and attention.

While a review of the literature on social justice leadership does not present a clear definition of social justice, there is a general framework for delineating it. Lee and McKerrow (2005) suggest that social justice is defined "not only by what it is but also by what it is not, namely injustice. By seeking justice, we anticipate the ideal. By questioning injustice, we approach it. Integrating both, we achieve it." (p. 1). These authors further assert that individuals for social justice seek to challenge political, economic and social structures that privilege some and disadvantage others. They challenge unequal power relationships based on gender, social class, race, ethnicity, religion, disability, sexual orientation, language, and other systems of oppression.

Each individual has the same indefeasible claim to a fully adequate scheme of equal basic liberties, which scheme is compatible with the same scheme of liberties for all; and social and economic inequalities are to be attached to offices and positions open to all under conditions of fair equality of opportunity. Research (e.g., Shapiro & Stefkovich, 2013) further assert that these inequalities are to be the greatest benefit of the least-advantaged members of society. In order to understand, promote, and enact social justice, school leaders must first develop a heightened and critical awareness of oppression, exclusion, marginalization, and justice. Awareness of social injustices, however, is not sufficient; school leaders must act when they identify inequity. As evidenced in the collection of international research by Bogotch (2014), these leaders are not only uniquely positioned to influence equitable educational practices, but their proactive involvement is also imperative.

DIGITAL INEQUALITY

Digital skills are an important aspect of ensuring that all young people are digitally included. Yet, there tends to be an assumption in popular discourse that young people can simply learn these skills by themselves. While experience of technologies forms an important part of the learning process, other resources (i.e., access to technology and support networks) plus clear motivations are required. Eynon and Geniets (2016) assert how poor access to technology, limited support networks and their current situation prevent young people from gaining the experiences they need to support the development of their digital skills; and how lack of experience and inadequate skills limit the extent to which they perceive the internet to be valuable in their lives. These researchers further assert that individual experiences, shaped very much by the wider social structure of which they are part, show how young people cannot simply be left to learn digital skills by themselves and that intervention is required to try to address some of the digital inequalities apparent in younger generations. This is part of the basic entitlement of every citizen, in every democracy in the world, to freedom of expression and the right to information. It is instrumental in building and sustaining democracy. Many policy makers and educator view these skills as critical to the creation of an equitable global 'Information Society' in which both developed and developing nations can share in social and economic development. Information leadership aims to develop *both* critical understanding *and* active participation.

DIGITAL EQUITY

Digital equity is about the "social justice goal" of "equitable access" and "effective use of technology for teaching and learning, access to content that is of high quality and culturally relevant" (Judge, Puckett, & Cabuk, 2004, p. 383). According to del Val (2006), there are several dimensions of digital equity that must be taken into consideration in order to help bridge the divide: content creation (i.e., opportunities for learners and educators to create their own content), effective use (i.e., educators must be skilled in using these resources effectively for teaching and learning), quality content (i.e., access to high quality digital content), cultural relevance (access to high quality, culturally relevant content), and technology resources. The literature pinpoints the importance of curriculum content in the integration of computer technology in schools (ILS—hardware, software, management, wiring, connectivity, etc.). Moreover professional development of teachers to utilize the latest technology in the classroom is crucial to the success of increased student access to computers in schools. Equity incorporates efficient and effective ILS utilization by teachers and students in schools. High quality, culturally relevant ILS in schools can help ensure that students have access to successful experiences as long as teachers keep pace with the acquisition of new knowledge and skills. Pedagogical content knowledge is essential with respect to integrating technology in the classroom effectively.

BRIDGING THE DIGITAL DIVIDE

Public schools can serve as the bridge to close the digital divide for students who do not have access to information technology in their homes. The advent of the internet in the mid-1990s increased the capacity of microcomputers to be used as effective tools to help students learn the necessary skills required in a changing job market. Students have a necessity, and some would say a right, to learn the computer skills demanded by the job market (Levy & Murname, 2004). At a time when more computers are made available in schools than ever before, the digital divide continues to widen and fewer people in the lowest SES groups are given the opportunity to join the world of computer technology and the internet.

As a result, the influence of leadership activity on institutional racism, gender discrimination, inequality of opportunity, inequity of educational processes, digital exclusion, and justice have gained currency and attention. The contributing national and international authors examine the digital divide in terms of social justice leadership, equity and access (Marshall & Oliva, 2006). It is within this context that the authors offer discussions from a lens of their choice, i.e. conceptual, review of literature, epistemological, etc. By adopting an educational approach to bridging the digital divide, researchers and practitioners can connect and extend long-established lines of conceptual and empirical inquiry aimed at improving organizational practices and thereby gain insights that might be otherwise overlooked, or assumed. This holds great promise for generating, refining, and testing theories of leadership for equity and access. It helps strengthen already vibrant lines of inquiry on social justice.

Crossing the Bridge of the Digital Divide: A Walk with Global Leaders is organized into a Foreword titled, *Bridging Divides in Cyber-Lives* written by Jabari Mahiri. This precedes 13 chapters divided into Parts I, II and III. Each part has a theme that further connects the chapters. Part I, *Dynamics of Digital and Social Inequity* has 3 chapters. In chapter 1, *Digital Equity and its Role in the Digital Divide*, authors Fortner, Normore and Brooks examine two decades of educational leadership and understanding of social justice. They examine how institutionalized theories, norms, and practices in schools and society lead to social, economic, and educational inequities, and how digital equity and its role in the digital divide, garnered increased attention. Chapter 2 is titled *An Examination of the Digital Divide and its Dividing Factors in Formal Educational Settings.* Ritzhaupt and Hohlfeld examine the factors used to characterize the Digital Divide in formal educational settings including, socio-economic status, education level, gender, age, geography, and race/ethnicity. In chapter 3, *The Digital Skills Paradox: How do Digitally Excluded Youth Develop Skills to Use the Internet?*, Eyon and Geniets demonstrate how poor access to technology, limited support networks and their current situation prevent young people from gaining the experiences they need to support the development of their digital skills; and how lack of experience and inadequate skills limit the extent to which they perceive the internet to be valuable in their lives.

Part II, *Issues of Digital Access* contains 7 chapters. In chapter 4, *Leading the Cohort across the Divide: Recent Best Practices to Enhance Cohort Teaching and Learning*, author Williams examines how a university program instituted a series of innovations to enhance and increase access, development and regular use of technology for teachers and school leaders. He explores the impact of the program in how it not only benefits these students, but the students in the schools they serve. In Chapter 5, *Walking the Pedagogical Line: Obstacles and Opportunities Transitioning to Digital and E-learning,* McLelland and Rintoul, explore and investigate the perceived obstacles and opportunities to learning, given that a greater number of courses at universities appear to be transitioning to an online format. They draw from an extensive literature of faculty perceptions related to online graduate learning as well as their own pedagogical insights. They recommend mindful consideration of both aspects of the obstacle/opportunity paradigm that confronts faculty teaching online graduate courses.

In chapter 6, *A Model for Addressing Adaptive Challenges by Merging Ideas: How One Program Designed a Hacking Framework to Address Adaptive Challenges and Discovered the Ecotone,* Zoller, Issa Lahera, and Jhun, argue that digital shifts and innovations occur with a rapidity that create a never-ending learning curve. These authors set the context for a new "hacking" leadership framework, and highlights the digital divide as the quintessential adaptive challenge for educators. Chapter 7, *Emerging Technologies for Learning: Using Open Education Resources (OER),* Caputo showcases innovative resources that have incorporated social computing technologies. He reveals some of the key implications for practice, and concludes with an outline of the current challenges faced by educators.

In chapter 8, *Partnering with Teachers to Bridge Digital Divides,* Zinger, Krishnan and Warschauer reiterate that teachers play a central role in student learning, so attending to teacher digital divides may also play an important role in bridging student digital divides. They present three cases of teachers who participated in professional development programs that partnered with the teachers and were designed to improve student learning through technology. Key findings found the importance of meeting teachers where they are instructionally both in terms of technology and their subject area. In chapter 9, *Social Networking*

Technology and the Social Justice Implications of Equitable Outcomes for First-Generation College Students, Fernandez, Deng and Zhao, discuss the findings from their qualitative study informed by social capital and social network theory. They examine the type of social ties and institutional resources that social media use enables FGCS to strengthen or build. The authors further underscore how the equitable access to technology has social justice implications when it comes to the post-secondary outcomes of first-generation college students. In chapter 10, *The Habitus and Technological Practices of Rural Students in South Africa: A Case Study*, Czerniewicz and Brown describe the habitus and technological practices of a South African rural student in his first year at university. This student is one of five self-declared rural students, from a group of 23 first-years in four South African universities, whose access to, and use of, technologies in their learning and everyday lives was investigated in 2011 using a 'digital ethnography' approach.

Part III contains 3 chapters focused on *Global Research and Development in Technology*. In chapter 11, *The Digital Divide in Scientific Development and Research: The Case of the Arab World*, author Salhi argues that a lack of an adequate scientific infrastructure will widen the digital divide between the Arab states and the advanced world further hindering the prospect of its human development and the realization of social justice. He uses empirical data to show the extent of integration of the scientific-knowledge in higher education and how diffused it is across the region. In chapter 12, *Assistive Technology for Students with Disabilities: An International and Intersectional Approach*, Kulkarni, Parmar, Selmi and Mendelson argue that augmentative and alternative communication devices tend to be promoted and discussed using a mostly dominant framing of race, culture, language and context. They call for a framework of intersectionality that embraces race, culture and language when supporting students with disabilities who utilize assistive technology devices. In chapter 13, *Online Resource Courses to Enhance Education Abroad Learning: The Digital & Enhanced International Learning Divide*, Rhodes and Raby assert that the mobility of university students from home countries to countries outside their home university campus continues to grow. They provide information about some of digital, e-learning, and social media tools and platforms that have recently been developed to support pre-departure, while-abroad, and re-entry learning for students who take part in international student mobility programs.

REFERENCES

Bogotch, I. E. (2014). Educational theory: The specific case of social justice as an educational leadership construct. In I. Bogotch & C. Shields (Eds.), *International handbook of educational leadership and social (in)justice* (pp. 51–65). Netherlands: Springer.

Carvin, A. (2006). *The gap: The digital divide network* (p. 70). Washington, DC: Reed Business Information.

del Val, R. E. (2006). Closing the equity gap. *Adult Education Quarterly, 57*(1), 90–91.

del Val, R. E., & Normore, A. H. (2008). Leadership for social justice: Bridging the digital divide. *University Council for Educational Administration (UCEA, International Journal of Urban Educational Leadership, 2,* 1–15. Available [On-line]: http://www.uc.edu/urbanleadership/current_issues.htm

Eynon, R., & Geniets, A. (2016). The digital skills paradox: How do digitally excluded youth develop skills to use the internet? *Learning, Media and Technology, 41*(3) 463–479.

Hohlfeld, T., Ritzhaupt, A. D., & Barron, A. E. (2013). Are gender differences in perceived and demonstrated technology literacy significant? It depends on the model. *Educational Technology Research and Development, 61*(4), 63–66.

Judge, S., Puckett, K., & Cabuk, B. (2004). Digital equity: New findings from the early childhood longitudinal study. *Journal of Research on Technology in Education, 36*(4), 383–396.

Lee, S. S., & McKerrow, K. (2005).Advancing social justice: Women's work. *Advancing Women in Leadership, 19,* 1–2.

Levy, F., & Murname, R. J. (2004). Education and the changing job market. *Educational Leadership, 62(*2), 80–83.

Marshall, C., & Oliva, O. (2006), *Leadership for social justice: Making revolutions in education.* Boston, MA: Pearson Education.

Murrell, P. J. (2006). Toward social justice in urban education: A model of collaborative cultural inquiry in urban schools. *Equity and Excellence in Education, 39,* 81–90.

Picciano, A. G. (2006). *Educational leadership and planning for technology* (4th ed.). Upper Saddle River, NJ: Pearson Education, Inc.

Pleasant, R. & Ritzhaupt, A. D. (2014). A review of video games and learning: Teaching and participatory culture in the digital age. *International Journal of Gaming and Computer-Mediated Simulations, 6*(1), 80—82.

Shapiro, J. P., & Stefkovich, J. A. (2013). *Ethical leadership and decision making in education: Applying theoretical perspectives to complex dilemmas* (4th ed.). Mahwah, NJ: Lawrence Erlbaum Associates, Inc.

SERIES EDITOR'S PREFACE

Jeffrey S. Brooks

I am pleased to serve as series editor for this book series, *Educational Leadership for Social Justice*, with Information Age Publishing. The idea for this series grew out of the work of a committed group of leadership for scholars associated with the American Educational Research Association's (AERA) Leadership for Social Justice Special Interest Group (LSJ SIG). This group existed for many years before being officially affiliated with AERA, and has benefitted greatly from the ongoing leadership, support, and counsel of Dr. Catherine Marshall (University of North Carolina-Chapel Hill). It is also important to acknowledge the contributions of the LSJ SIG's first Chair, Dr. Ernestine Enomoto (University of Hawaii at Manoa), whose wisdom, stewardship, and guidance helped ease a transition into AERA's more formal organizational structures. This organizational change was at times difficult to reconcile with scholars who largely identified as non-traditional thinkers and push toward innovation rather than accept the status quo. As the second Chair of the LSJ SIG, I appreciate all of Ernestine's hard work and friendship. Moreover, I also thank Drs. Gaetane Jean-Marie and Whitney Sherman Newcomb, the third and fourth Chairs of the LSJ SIG for their visionary leadership, steadfast commitment to high standards and collaborative scholarship and friendship.

I am particularly indebted to my colleagues on the LSJ SIG's first Publications Committee, which I chaired from 2005–2007: Dr. Denise Armstrong, Brock University; Dr. Ira Bogotch, Florida Atlantic University; Dr. Sandra Harris, Lamar

Crossing the Bridge of the Digital Divide: A Walk with Global Leaders, pages xxi–xxiii.
Copyright © 2019 by Information Age Publishing

University; Dr. Whitney Sherman, Virginia Commonwealth University, and; Dr. George Theoharis, Syracuse University. This committee was a joy to work with and I am pleased we have found many more ways to collaborate—now as my fellow Series Editors of this book series—as we seek to provide publication opportunities for scholarship in the area of leadership for social justice.

We also owe a debt of gratitude to George Johnson, our publisher at Information Age Publishing, who has always been a great supporter of the series.

This book, *Crossing the Bridge of the Digital Divide: A Walk with Global Leaders*, co-edited by Anthony H. Normore and Antonia Issa Lahera explores the combined effect of the rapid growth of information as an increasingly fragmented information base, a large component of which is available only to people with money and/or acceptable institutional affiliations. The book is global in scope, with chapters that focus on issues in the Unites States, Canada, South Africa, New Zealand, and the UK. Collectively considered, authors argue that education and leadership have great potential for promoting equity in technology education.

Again, welcome to this twenty-third book in this Information Age Publishing series, *Educational Leadership for Social Justice*. You can learn more about the series at our web site: http://www.infoagepub.com/series/Educational-Leadership-for-Social-Justice. I invite you to contribute your own work on equity and influence to the series. We look forward to you joining the conversation.

—*Jeffrey S. Brooks*
Royal Melbourne Institute of Technology
RMIT University, Australia

OTHER BOOKS IN THE EDUCATIONAL LEADERSHIP FOR SOCIAL JUSTICE BOOK SERIES

Anthony H. Normore, Editor (2008). *Leadership for social justice: Promoting equity and excellence through inquiry and reflective practice.*

Tema Okun (2010). *The emperor has no clothes: Teaching about race and racism to people who don't want to know.*[1]

Autumn K. Cypres & Christa Boske, Editors (2010), *Bridge leadership: Connecting educational leadership and social justice to improve schools.*

Dymaneke D. Mitchell (2012). *Crises of identifying: Negotiating and mediating race, gender, and disability within family and schools.*

Cynthia Gerstl-Pepin & Judith A. Aiken (2012). *Defining social justice leadership in a global context: The changing face of educational supervision.*

Brian D. Fitch & Anthony H. Normore, Editors. (2012). *Education-based incarceration and recidivism: The ultimate social justice crime-fighting tool.*

Christa Boske, Editor (2012). *Educational leadership: Building bridges between ideas, schools and nations.*

Jo Bennett (2012). *Profiles of care: At the intersection of social justice, leadership, & the ethic of care.*

---

[1] Winner of the American Educational Studies Association 2011 Critics Choice Award

Elizabeth Murakami-Ramalho & Anita Pankake (2012). *Educational leaders encouraging the intellectual and professional capacity of others: A social justice agenda.*

Jeffrey S. Brooks & Noelle Witherspoon Arnold, Editors (2013). *Antiracist school leadership: Toward equity in education for America's students.*

Jeffrey S. Brooks & Noelle Witherspoon Arnold, Editors. (2013). *Confronting racism in higher education: Problems and possibilities for fighting ignorance, bigotry and isolation.*

Mary Green (2014). *Caring leadership in turbulent times: Tackling neoliberal education reform.*

Carol Mullen (2014). *Shifting to fit: The politics of black and white identity in school leadership.*

Anthony H. Normore & Jeffrey S. Brooks, Editors (2014). *Educational leadership for ethics and social justice: Views from the social sciences.*

Whitney N. Sherman & Katherine Mansfield, Editors (2014). *Women interrupting, disrupting, and revolutionizing educational policy and practice.*

Carlos McCray & Floyd Beachum (2014). *School leadership in a diverse society: helping schools to prepare all students for success.*

M. C. Kate Esposito & Anthony H. Normore, Editors (2015). *Inclusive practices for special populations in urban settings: The need for social justice leadership.*

Jeffrey S. Brooks & Melanie C. Brooks, Editors (2015). *Urban educational leadership for social justice: International perspectives.*

Natasha Croom & Tyson Marsh (2016). *Envisioning Critical Race Praxis in Higher Education through Counter-Storytelling.*

Tyson Marsh & Natasha Croom (2016). *Envisioning a Critical Race Praxis in K-12 Leadership through Counter-Storytelling.*

William DeJean & Jeff Sapp (2016). *Dear Gay, Lesbian, Bisexual, and Transgender Teacher: Letters of Advice to Help You Find Your Way.*

Hoaihuong Nguyen & Jeanne Sesky (2017). *Within Reach: Providing Universal Access to the Four Pillars of Literacy.*

PART I

DYNAMICS OF DIGITAL AND SOCIAL INEQUITY

CHAPTER 1

DIGITAL EQUITY AND ITS ROLE IN THE DIGITAL DIVIDE

Kitty Fortner, Anthony H. Normore, and Jeffrey S. Brooks

Over the past two decades, educational leadership scholars have made significant contributions to our understanding of social justice (Adams, Bell, & Griffin, 1997; Applebaum, 2009; Bigelow, Christensen, Karp, Miner, & Peterson, 1994; Marshall & Oliva, 2006; Marshall & Ward, 2004). Important conceptual research suggests that a social justice orientation to educational leadership practice and research can address "how institutionalized theories, norms, and practices in schools and society lead to social, economic, and educational inequities" (Dantley & Tillman, 2006, p. 17). In response to a more equitable practice in urban schools, this chapter explores digital equity and its role in the digital divide, which have both garnered increased attention (Normore & Brooks, 2014; Marshall & Oliva, 2006).

Access to technology, specifically computers and the internet continues to plague US schools, specifically low SES schools. The rapid growth of information, the limited and often inequitable distribution of resources and the continued disproportionality of opportunities between schools, especially when technology is concerned, is most evident for schools serving children of color (Ritzhaupt, Dawson, & Cavanaugh, 2012). The digital divide addresses issues concerning equal opportunity, equity and access, and speaks directly to the negative effect on the development of marginalized and otherwise disadvantaged students in education systems (de Val & Normore, 2008; Selwyn, Gorard, & Williams, 2001).

Previous research on the digital divide focused on the limited access to technology and the internet, which yielded students who were less prepared for the increasingly global market that is emerging in the 21st century. More resent research indicates the issue of access has decreases with the inclusion of mobile devices, cell phones and mini computers (Fairlie, 2017), however the issue of use continues at a critical level. Research (e.g., Hatlevik, 2013; Howard, Busch, & Sheets, 2010; Philip, Cottrill, Farrington, Williams, & Ashmore, 2017; Piacciano, 2007; Tansley, 2006) clearly indicates that this is a global phenomenon that has caused a widening equity gap in primary, secondary and higher education across all continents. Consequently, information leadership has become more critical than ever as discourses about the knowledge economy focus on the necessity of educating ALL students with skills for the global workplace.

School leaders and those who prepare them will need to know why, when, and how to use all of these tools and think critically about the information they provide. To do so will enable educators to interpret and make informed judgments as users of information sources. It will also enable them to become producers of information in their own right, and thereby become more powerful participants in society. This is part of the basic entitlement of every citizen, in every democracy in the world, to freedom of expression and the right to information (Abdelaziz, 2004). It is instrumental in building and sustaining democracy. Many policy makers and educators view these skills as critical to the creation of an equitable global 'Information Society' in which both developed and developing nations can share in social and economic development. Information leadership aims to develop both critical understanding and active participation

LEARNING ABOUT THE "DIGITAL DIVIDE'

The term "digital divide" is polysemous and became a part of the educator's vernacular in the mid-1990s. Traditionally, the term generally describes social inequity between individuals who have and do not have access to information and communication technology (van Dijk, 2006). Although, many researchers have asserted that education continues to reflect a "digital divide" between the information "haves" and "have nots" in society along racial and socio-economic factors (e.g., Kaiser Family Foundation, 2004; Tansley, 2006) the recent surge of computers, smart devices and internet access in public schools as an integrated effort to infuse information technology into the curriculum has expanded the dialogue from the "haves" and "have nots" to the "cans' and "cannots"(Dolan, 2016; Rogers, 2016; Rowsell, Morrell, & Alvermann, 2017). Ritzhaupt, Liu, Dawson, and Barron (2013) use a multileveled approach when defining the digital divide which includes both the "haves" and the "have nots" as well as the "cans" and "cannots": (1) equitable access to hardware, software, the Internet, and technology support within schools; (2) the use of technology (frequency and purpose) by students and teachers in the classroom; and (3) whether student users know how to use information and communication technology for their personal empowerment.

Nearly 100% of US public schools provide access to the internet for students and school with the highest poverty concentration even provided before and after school access to computers with internet access for student use. However, the issues concerning equal educational opportunity and equity that have an effect on the development of disadvantaged students in public schools is encompassed in both the access at home and the utilization at home and in school of technology for personal empowerment (Gorski, 2009; Instefjord & Munthe, 2017; Ritzhaupt et al., 2013).

Disadvantaged students are usually the ones who do not have access to computer technology outside of their schools, resulting in this digital divide where minority students and the poor living in urban and rural areas become less prepared for the increasingly competitive global market that is emerging in the 21st century. Occupations utilizing computers and information technology is projected to grow 13 percent from 2016 to 2026, faster than the average for all occupations (U. S. Bureau of Labor, 2016). With these statistics at the nexus of public school, one pervasive issue that needs to be addressed is whether public schools can serve to bridge the widening gap of the digital divide. The digital divide is not limited to the U.S. It is a global phenomenon that encompasses a widening equity gap in primary, secondary, and higher education (Cohron, 2015; Gudmundsdottir, 2010; Hatlevik, 2013; Philip et al., 2017)

Over the past two decades, educational leadership scholars have made significant contributions to our understanding of social justice (Adams, Bell, Goodman, & Joshi, 2016; Applebaum, 2009; Bigelow, Karp, & Au, 2007; Delpit, 2006; Giroux, 1994; Marshall & Oliva, 2006; Marshall & Ward, 2004; Scanlan, 2013). Important conceptual research suggests that a social justice orientation to educational leadership practice and research can address "how institutionalized theories, norms, and practices in schools and society lead to social, economic, and educational inequities" (Dantley & Tillman, 2010, p. 18). In response to a more equitable practice in urban schools, this research explored social justice leadership and the digital divide, which have both garnered increased attention (Education Reform Network, 2003; Marshall & Oliva, 2006).

According to Marshall and Oliva (2006), finding conceptual inspiration and guidance in notions of equity and equality, researchers and practitioners have begun to develop a pedagogy of leadership based on an ethic of care. Furthermore, they have embraced the moral imperative of improving "practice and student outcomes for minority, economically disadvantaged, female, gay/lesbian, and other students who have not traditionally been served well in schools" (Marshall & Oliva, 2006, p. 7). As a result, the influence of leadership activity on institutional racism, gender discrimination, inequality of opportunity, inequity of educational processes, and justice have gained currency and attention. According to Rawls (2001) each individual has the same indefeasible claim to a fully adequate scheme of equal basic liberties, which scheme is compatible with the same scheme of liberties for all; and social and economic inequalities are to be attached to offices and positions open to all under conditions of fair equality of opportunity. He further asserts that these inequalities are to be the greatest benefit of the least-advantaged members of society (p. 291).

The unjust reality of the world has been explained with concepts such as hegemony (Gramsci, 1975; O'brien & Williams, 2016), a culture of power (Delpit, 2006), an interrelationship of cultural inequalities in cultural politics (Giroux, 1994), racial marginalization (Ladson-Billings, 2012). In order to understand, promote, and enact social justice, school leaders must first develop a heightened and critical awareness of oppression, exclusion, marginalization, and justice. Awareness of social injustices, however, is not sufficient; school leaders must act when they identify inequity. Given the unique position of school leaders to influence equitable educational practices, their proactive involvement is imperative. As Larson and Murtadha (2002) note, "throughout history, creating greater social justice in society and in its institutions has required the commitment of dedicated leaders" (p. 135). While a review of the literature on social justice leadership does not present a clear definition of social justice, there is a general framework for delineating it. Lee and McKerrow (2005, p. 1) suggest that social justice is defined "not only by what it is but also by what it is not, namely injustice. By seeking justice, we anticipate the ideal. By questioning injustice, we approach it. Integrating both, we achieve it." These authors further assert that individuals for social justice seek to challenge political, economic and social structures that privilege some and disadvantage others. They challenge unequal power relationships based on gender, social class, race, ethnicity, religion, disability, sexual orientation, language, and other systems of oppression.

TECHNOLOGY "HAVES" AND "HAVE NOTS"

Literature continues indicates that poor disenfranchised minority students have limited opportunities to prepare for the economic demands of the 21st century (e.g., Freeman, 2005; Ritzhaupt et al., 2013; Welner & Weitzman, 2005). The issue of access referred to having the physical components of computers (hardware and software) in the classrooms or somewhere in the schools (e.g. media center). Today, Ninety-two percent of teenagers aged 12–17 years go online daily (in school or at home), while 97% of them play computer, web portal, or console games and 75% of them own a smartphone (Lenhart, 2015). It may seem that 30 years later there are more students than ever before with access to computers in schools and at home; however, the access and use of computers in schools and at home is not equal for all students (Ritzhaupt et al., 2013). Equal access to information technology continues to be a struggle for poor and other minority students in the U.S.

Today, the general population has more choices and uses for information and communication technologies than ever before. Mobile computing and reduced prices has afforded many access to information and communication technologies. According to a Pew Research Center 2016 study, 88% of American teens ages 13 to 17 have or have access to some type of computing device. The study continues on to say 87% have or have to access to a desktop/laptop computer, 81% to a gaming console, 75% a smartphone, 58% a tablet and 15% a basic cell phone (Rainie, 2016). Common Sense Media (2017)found that 98% of children aged 0–8 had used a mobile device to play games, use apps, or watch videos. The U.S. Department of Commerce, in a document written by the National Telecommunications and Information Administration (NTIA), called the United States a "Digital Nation" (NTIA, 2014) reporting that home access to computers, the Internet, and broadband services had increased between 1997–2012, revealing the majority of homes have access to a device with an Internet connection. Access to technology has increased however effective utilization in schools continues as an issue for schools. Scholars are calling for a broader definition of digital divide that encompasses issues of equity in students' access and successful utilization of computer technology in schools.

DIGITAL ACCESS AND EQUITY

Having access to information and communication technology continues to be the starting point, moreover, current literature examines equity in terms of how access is supported or constrained by technological and social factors for diverse groups of youth. As previously stated 88% of youth have access to a mobile phone however 15% have only a basic phone which leaves 12% with no phone at all. This leaves 27% who are unable to participate at a rate that leverages learning. The predominate factor for those without access is cost. Participants in the 2106 Pew Research Center survey on the digital divide found 33% of participants stated that monthly subscription cost for broadband services were too high while 10% stated the cost of computers were too expensive (Raine, 2016). Although students may have access at school, the limited access at home to the internet and smart devices creates inadequate and restricted learning for disadvantaged youth (Martin, 2016). Digital equity is defined contextually to address fair access to information technology for children. Moreover, the content and use of the hardware and software are crucial to the optimal use of computer technology in teaching and learning. Mobile access does not mean that youth have the necessary skills to utilize technology in a manner that creates engaged and connected learning experiences (Adams et al., 2016; Bennett & Maton, 2010; Fairlie, 2017; Martin, 2016).

Technology in the home is another factor that affects the education of students in school (Bulman & Fairlie, 2016; Dolan, 2016; NTIA, 2011). Although the internet is available to everyone, individuals living in poverty are more likely to have lower levels of computer skills and computer access, therefore lagging in the opportunities that both computers and the internet provide. Opportunity for all? Technology and learning in lower-income families, a report published in 2016 by the Joan Ganz Cooney Center at Sesame Workshop reports that 94% of families living in the low- and moderate-income levels have some form of Internet access (computer with internet connection at home or smart mobile device with a data plan). The majority of these families (56%) rely on mobile only internet access with issues such as low data limits on plan, loss of service due to inability to make payment, and too many people sharing the same phone. The families who have a computer at home reported slow access, no access due to unpaid access fees, and too many people using the same computer (Rideout & Katz, 2016). Poorer students are less prepared as they enter school and lag behind other students from higher SES groups. The literature suggests students without Internet access at home are less likely to use the internet for information gathering about things that they are interested in: 35% of those with mobile-only access compared to 52% of those with home access (Rideout & Katz, 2016; Rowsell et al., 2017). Additionally approximately half of Black and Latino children have access to a home computer, compared to 85.5% of White children (Rainie, 2016).

Public schools can serve as the bridge to close the digital divide for students who do not have access to information technology in their homes. The advent of the internet in the mid-1990s increased the capacity of microcomputers to be used as effective tools to help students learn the necessary skills required in a changing job market. With the declining cost of K-12 internet access, an estimated 97% of public schools are now connected to the internet compared to 30% of public schools in 2013 (Marwell, 2017). These figures can be misleading as it misrepresents the access to technology experienced by poorer and disadvantaged students. Those who have more access include teachers, administrators, and office personnel in lieu of students. According to the Education Super Highway 2017 State of the States Report, 6.5 million students are still in need internet access, 10,000 schools have less than sufficient bandwidth for classroom connectivity which limits student's ability to adequately prepare for college and careers and limits teacher's instruction (Marwell, 2017).

THE GAP IN THE DIGITAL DIVIDE

Internet access between whites and minorities (Hispanics, Blacks) differ considerably in homes as well as in schools (del Val, 2006). Vigdor and Ladd reported that 90% of white students in North Carolina public schools have computers at home and only 75% of black students had computers at home (2010). US Census Bureau (2013) reports 85% of non-Hispanic Whites had home computers with 77% connected to the internet, compared to 75% of Black with 61% internet access, and 80% Hispanic with 68% internet access; and Asians surpass Whites in having home computers by 7% (File & Ryan, 2014). Other studies show that there was a disparity between children in highest SES groups having more access to computers and the internet than children in lowest SES groups (Dolan, 2016; Ritzhaupt et al., 2013). The literature indicates that the digital divide appears at all levels of schooling even the parents level of education (Rideout & Katz, 2016).

The digital divide is a term coined by former Present Clinton Former President Bill Clinton in the 1996 speech in Knoxville, Tennessee. The term refers to the gap that exist between people with access to digital and information technology and those who do not have access. It describes the inability to access computer technology and the use of the internet by certain segments of society (i.e. the poorest) from participation "in the global information society" (Cohron, 2015). The multileveled approach of the digital divide provided by Ritzhaupt, Liu, Dawson, and Barron (2013) focuses on: (1) equitable access to hardware, software, the Internet, and technology support within schools; (2) the use of technology (frequency and purpose) by students and teachers in the classroom; and (3) whether student users know how to use information and communication. This approach moves the discussion on the digital divide beyond looking at the physical possession of a computer, smart device and access to the internet to the knowledge and skills needed to utilize technology for effective growth.

TECHNOLOGY "CANS AND CANNOTS"

Digital equity is about the "social justice goal" of "equitable access" and "effective use of technology for teaching and learning, access to content that is of high quality and culturally relevant" (Judge, Puckett, & Cabuk, 2004, p. 383). Although there remains a group of student with no access or limited access to computers or the internet, the efficient use of technology for those with access is also an issue to be addressed. In recent discussion of the digital divide, a growing concern over the "chasm between students' out-of-school and in-school uses of technology, (Dolan, 2016), the lack of teacher confidence and capacity concerning the integration of technology in to the classrooms (Wang, Hsu, Campbell, Coster, & Longhurst, 2014) and security concerns surrounding computer and internet usage (Charania & Davis, 2016; Minneapolis, 2014) have emerged. The "cans" are students who are connected at home and in school. They not only own technology but they utilize it in a manner that demonstrates proficiency. They are tech savvy students who actively produce, create, design, and publish online. They leverage technology utilizing multiple aspects to promote their learning and understanding of the world.

The "cannots" have fewer opportunities to actively leverage technology. With limited access in home and at school, these students typically attend low SES schools with limited access to computers, insufficient bandwidth, low budgets, and teacher who lack adequate technology training. These students do not have the opportunity to utilize technology in a way that is creative or tied to their interest (Braun, Hartman, Hughes-Hassell, & Kumasi, 2014; Martin, 2016). Often their technology usage is limited to the technology abilities of teachers who are in need of training (Dolan, 2016; Khalifa, Gooden, & Davis, 2016; Rideout & Katz, 2016) and consist of repetitive practice rather than more sophisticated, intellectually complex applications. Teacher play an important role in shaping students' technology experiences and districts and school leaders need to provide both teacher support and platforms for student access and internet usage. This lack of access and inability to utilize technology is a challenge to digital equity. The term "opportunity gap"(Carter & Welner, 2013) places the onus of this equity challenge on educators, especially educational leaders, to address equity issue by consciously providing more opportunities for students to achieve. According to Kuntz (2015) educational inequities need the combined ability, intelligence, commitment, and corporation of scholars, leaders, and practitioners within the field of education to challenge the status quo of dominant hegemonic discourses.

USING ETHICAL PARADIGMS TO ADDRESS THE DIGITAL DIVIDE

A theoretical framework for applying ethics using multiple paradigms is suggested here for the purpose of addressing the issue of equity and social justice in the information age of computers. These paradigms consist of the ethics of justice, critique, care, profession (Shapiro & Stefkovich, 2016), and community (Furman, 2004). The ethic of justice highlights the right of equal educational opportunity that is of high quality available to all students (fairness, equity, equality) whereas the ethic of critique examines the current policies and practices that may perpetuate inequalities inherent in public schools (bureaucracy). As changes in policy and practice lead to digital equity in providing equal access to computer technology and its effective use, then disadvantaged students would not be affected by race/ethnicity or SES in learning computer skills in order to succeed in school and (later on) in the work force. The ethic of care ensures that children are put first and education serves the best interest of students. As these paradigms merge, the ethic of the profession acts as a call to action for educational leaders to plan, implement, and sustain efficient and effective ILS in their schools. The ethic of community addresses the greater good for the greatest number in supporting excellence in education and digital equity for all students in public schools.

These paradigms support social justice education—in particular the work of (Freire, 1970), Smith-Maddox and Solórzano (2002), Rogers, Morrell, and Enyedy (2007), Kuntz (2015), and Irizarry (2015) who asserts that we must engage in problem-posing methodology in order to identify and name inequities, analyze the cause of the inequities, and find solutions through socially and culturally situated, intentional truth telling. The idea of digital equity or connection between the digital divide and social justice begins with providing equitable and high quality education for all students (Resta & Laferrière, 2015). Equitable access to technology, effective use of technology for teaching and learning, access to high quality content, and the opportunity to create new content are all dimensions of digital equity (Judge, Puckett, & Cabuk, 2004). Gillian and Ward (2004) claim that social justice education is a "belief in our own humanity and the power to assert our moral authority in the face of continuing injustice and intolerance" (p. 69). Moreover, technology integration in public schools must be relevant and meaningful to students and teachers (Charania & Davis, 2016; Ritzhaupt et al., 2013), contain culturally responsive content (Furman & Sheilds, 2005) and confront digital inequity by establishing new educational policy structures (Resta & Laferrière, 2015).

According to the Education Reform Network (2003) there are several dimensions of digital equity that must be taken into consideration in order to help bridge the divide: content creation (i.e., opportunities for learners and educators to create their own content), effective use (i.e., educators must be skilled in using these resources effectively for teaching and learning), quality content (i.e., access to high quality digital content), cultural relevance (access to high quality, culturally relevant content), and technology resources. The literature pinpoints the importance of pedagogical content knowledge in the effective integration of technology in schools (Charania & Davis, 2016; Dolan, 2016). Moreover professional development that builds digital competence in classroom teachers and school administrators is crucial to the success and increased student access and usage of technology in the classroom (Gamrat, Zimmerman, Dudek, & Peck, 2014; Miranda & Russell, 2012; Tondeur et al., 2017).

High quality, culturally relevant technology utilization in schools can help ensure that students have access to successful experiences as long as teachers keep pace with the acquisition of new knowledge and skills (Resta & Laferrière, 2015). Equity incorporates efficient and effective technology utilization by teachers and students in schools. To address the issues of digital equity, school leaders are in a position to ensure opportunities are made available for all students in schools. Social justice is constructed in relation to experiential knowledge of social injustice and the truth telling that surrounds that knowledge. School leaders much work to address and eliminate injustices in schools. Theoharis (2007) defines social justice leadership in education as principals with advocacy agendas, leadership practice and vision that include "issues of race, class, gender, disability, sexual orientation, and other historically and currently marginalizing conditions in the United States" (p. 223). He continues on to say that these principals maintain a vision of equity and justice while committing their energy to the creation of a more democratic and empowered staff (Theoharis, 2010) . At a time when criticisms are being voiced about the eroding ethics of society, it becomes vital that decisions and actions for 21st century educational leaders be based on ethical and moral foundations. Successful social justice leadership will involve moral choices with an emphasis on sense and meaning, morality, self-sacrifice, duty, and obligation. Recent research reflects a distinct trend emphasizing that effective school leaders advocate for social justice and maintain an ethical orientation (Bogotch, 2005; Furman & Sheilds, 2005; Lee & McKerrow, 2005; Normore & Blanco, 2006). According to Jazzar and Algozzine (2007), in order for school and district leaders to be successful in the new millennium, "the pendulum must swing back to values and moral dimension" (p. 155). Brown and Treviño (2006) offer traits on how transformational leaders employ ethical behavior by 1) demonstrating a genuine care for others; 2) acting with integrity aligning their behavior with moral principles; 3) considering the ethical consequences of their decisions; and 4) being ethical role models for others. Balyer (2012) discuss how principals use transformational leadership as a frame for their attitudes to move their schools forward. She outlines four major characteristics, 1) idealized influence which refers to the leader's behavior in relation to their followers, 2) inspirational motivation refers to methods used motivate and inspire those around them, 3) individualized consideration refers the way the leader treats each individual, 4) intellectual stimulation refers to the leader's effort to stimulate followers to be innovative and creative. School leaders who lead for social justice place student and staff as priorities and advocate for the needs of both.

CONCLUSION

In this paper, we took a closer look at the digital divide and its widening and narrowing taking place in education. Realizing the fact that although education has made great strides in move towards equity in access, equity in the use of technology continues as a critical issue. Understanding that school leaders play an important role in addressing this issue, use of a multiple paradigm approach as outlined by Shapiro and Stefkovich (2016) may help to revolutionize the field of educational leadership to successfully bridge the achievement gap and meet the challenges of the 21st century.

Shapiro and Stefkovich (2016) discuss the ethics of justice, critique, and care, in combination with the ethic of the profession providing a frame for these forces to complement each other and assist the school leader in their ability to make appropriate decisions. This process, serves the "best interests of the student" and becomes the "moral imperative" (p. 25). When moral authority overcomes bureaucratic leadership in a school, the outcomes can be extraordinary for all students.

REFERENCES

Abdelaziz, A. (2004). Information competency for lifelong learning. In *World Library and Information Congress: 70th IFLA General Conference and Council* (p. 4). Retrieved from http://www.ifla.org/IV/ifla70/papers/116e-Abid.pdf

Adams, M., Bell, L. A., Goodman, D., & Joshi, K. J. (2016). *Teaching for diversity and social justice.* (M. Adams & L. A. Bell, Eds.) (3rd ed.). New York, NY: Routledge.

Adams, M., Bell L.A., & Griffin, P. (Eds.). (1997). *Teaching for diversity and social justice: A Sourcebook.* New York, NY: Routledge.

Applebaum, B. (2009). Is teaching for social justice a " liberal bias "? *The Teachers College Record, 111*(2), 376–408.

Balyer, A. (2012). Transformational leadership behaviors of school principals: A qualitative research based on teachers ' perceptions. *International Online Journal of Educational Sciences, 4*(3), 581–591.

Bennett, S., & Maton, K. (2010). Beyond the "digital natives" debate: Towards a more nuanced understanding of students' technology experiences. *Journal of Computer Assisted Learning, 26*(5), 321–331.

Bigelow, B., Christensen, L., Karp, S., Miner, B., & Peterson, B. (1994). *Rethinking our classrooms: Teaching for equity and justice.* Milwaukee, WI: Rethinking Schools, Ltd.

Bigelow, W., Karp, S., & Au, W. (2007). *Rethinking our classroom: Teaching for equity and justice* (New Edition). Milwaukee, WI: A Rethinking Schools Publication.

Bogotch, I. E. (2005). *Social justice as an educational construct: Problems and possibilities.* Paper presented at the annual meeting of the University Council of Educational Administration. Nashville, TN, November, 2005.

Braun, L. W., Hartman, M. L., Hughes-Hassell, S., & Kumasi, K. (2014). The future of library services for and with teens: A call to action. *Library and Information Services Journal, 59.* Retrieved from http://www.ala.org/yaforum/sites/ala.org.yaforum/files/content/YALSA_nationalforum_Final_web_0.pdf

Brown, M. E., & Treviño, L. K. (2006). Ethical leadership: A review and future directions. *Leadership Quarterly, 17*(6), 595–616. https://doi.org/10.1016/j.leaqua.2006.10.004

Bulman, G., & Fairlie, R. W. (2016). Chapter 5—Technology and education: Computers, software, and the internet. *Handbook of the Economics of Education, 5,* 239–280. https://doi.org/10.1016/B978-0-444-63459-7.00005-1

Carter, P. L., & Welner, K. G. (Eds.). (2013). *Closing the opportunity gap: What America must do to give every child an even chance.* New York, NY: Oxford University Press.

Charania, A., & Davis, N. (2016). A smart partnership: Integrating educational technology for underserved children in India. *Educational Technology and Society, 19*(3), 99–109.

City of Minneapolis. (2014). *"Digital divide" survey points to digital equity opportunities.* Washington, DC: US Fed News Service.

Cohron, M. (2015). The continuing digital divide in the United States. *Serials Librarian, 69*(1), 77–86. https://doi.org/10.1080/0361526X.2015.1036195

Common Sense Media. (2017). *The common sense census: Media use by kids age zero to eight 2017.* San Francisco, CA.

Dantley, M. E., & Tillman, L. C. (2006). Social justice and moral transformative leadership. In C. Marshall & M. Oliva (Eds.), *Leadership for social justice: Making revolutions in education.* (pp. 16–30). New York, NY: Pearson.

Dantley, M. E., & Tillman, L. C. (2010). Social justice and moral transformative leadership. In M. Catherine & M. Oliva (Eds.), *Leadership for social justice: Making revolutions in education* (2nd ed., pp. 19–34). San Francisco, CA: Pearson Education, Inc.

del Val, R. E. (2006). Book review: *Closing the equity gap* (2005, Geoff Layer, Ed.). *Adult Education Quarterly, 57*(1), 90–91.

de Val, R. E., & Normore, A. (2008). Leadership for social justice: Bridging the digital divide. *University Council for Educational Administration (UCEA, International Journal of Urban Educational Leadership, 2,* 1–15. Retrieved from http://www.uc.edu/urbanleadership/current_issues.htm

Delpit, L. (2006). *Other people's children: Cultural conflict in the classroom.* New York, NY: The New Press.

Dolan, J. E. (2016). Splicing the divide: A review of research on the evolving digital divide among K–12 students. *Journal of Research on Technology in Education, 48*(1), 16–37. https://doi.org/10.1080/15391523.2015.1103147

Education Reform Network. (2003). *The five dimensions of digital equity.* Retrieved from http://digitalequity.edreform.net/

Fairlie, R. W. (2017). Have we finally bridged the digital divide? Smart phone and Internet use patterns race and ethnicity. *First Monday*, *22*(9), 1–11. https://doi.org/10.5210/fm.v22i19.7919

File, T., & Ryan, C. (2014). Computer and Internet use in the United States. *American Community Survey Reports*, (November), 16. https://doi.org/10.15511/tahd.16.21672

Freeman, E. (2005). No child left behind and the denigration of race. *Equity and Excellence in Education*, *38*(3), 190–199. https://doi.org/10.1080/10665680591002560

Freire, P. (1970). *Pedagogy of the oppressed.* New York, NY: Continuum.

Furman, G. C. (2004). The ethic of community. *Journal of Educational Administration*, *42*(2), 215–235. https://doi.org/10.1108/09578230410525612

Furman, G. C., & Sheilds, C. M. (2005). How can educational leaders promote and support social justice and democratic community in schools? In W. A. Firestone & C. Riehl (Eds.), *A new agenda for educational leadership* (pp. 119–137). New York, NY: Teachers College Press.

Gamrat, C., Zimmerman, H. T., Dudek, J., & Peck, K. (2014). Personalized workplace learning: An exploratory study on digital badging within a teacher professional development program. *British Journal of Educational Technology*, *45*(6), 1136–1148. https://doi.org/10.1111/bjet.12200

Gillian, C., & Ward, J. (2004). Forward. In V. Siddle Walker & J.R. Snarey (Eds.), *Racing moral formation: African American perspectives on care and justice* (pp. ix–xii). New York, NY: Teachers College Press.

Giroux, H. A. (1994). Teachers, public life and curriculum reform. *Peabody Journal of Education*, *69*(3), 35–47.

Gorski, P. C. (2009). Insisting on digital equity. *Urban Education*, *44*(3), 348–364. https://doi.org/10.1177/0042085908318712

Gramsci, A. (1975). *Selections from the prison notebook* (Q. Hoare & G. N. Smith, Trans., Eds.). New York, NY: International Publishers.

Gudmundsdottir, G. (2010). From digital divide to digital equity: Learners' ICT competence in four primary schools in Cape Town, South Africa. *International Journal of Education and Development Using ICT*, *6*(2), 84–105. Retrieved from http://www.editlib.org/p/42335/

Hatlevik, O. E. (2013). An emerging digital divide in urban school children's information literacy: Challenging equity in the Norwegian school system. *First Monday*, *18*(4), 1–12.

Howard, P. N., Busch, L., & Sheets, P. (2010). Comparing digital divides : Internet access and social inequality in Canada and the United States. *Canadian Journal of Communication*, *35*(1), 109–128. Retrieved from http://cjc-online.ca/index.php/journal/article/viewFile/2192/2161

Instefjord, E. J., & Munthe, E. (2017). Educating digitally competent teachers: A study of integration of professional digital competence in teacher education. *Teaching and Teacher Education*, *67*, 37–45. https://doi.org/10.1016/j.tate.2017.05.016

Irizarry, J. (2015). *Latinization of US schools: Successful teaching and learning in shifting cultureal contexts.* New York, NY: Routledge.

Jazzar, M., & Algozzine, R. (2007). *Keys to 21st century educational leadership.* Boston, MA: Pearson Education/Allyn & Bacon.

Judge, S., Puckett, K., & Cabuk, B. (2004). Digital equity: New findings from the early childhood longitudinal Study. *Journal of Research on Technology in Education*,

36(4), 383–396. Retrieved from http://www.tandfonline.com/doi/abs/10.1080/153 91523.2004.10782421

Kaiser Family Foundation. (2004). *The digital divide. Survey Snapshot, August.* Washington, D.C. Retrieved from www.kff.org

Khalifa, M. A., Gooden, M. A., & Davis, J. E. (2016). Culturally responsive school leadership: A synthesis of the literature. *Review of Educational Research, 86*(4), 1272–1311. https://doi.org/10.3102/0034654316630383

Kuntz, A. M. (2015). *The responsible methodologist: Inquiry, truth-telling, and social justice.* Walnut Creek, CA: Left Coast Press.

Ladson-Billings, G. (2012). Through a glass darkly: The persistence of race in education research & scholarship. *Educational Researcher, 41*(4), 115–120. https://doi.org/10.3102/0013189X12440743

Larson, C. L., & Murtadha, K. (2002). Leadership for social justice. *Yearbook of the National Society for the Study of Education, 101*(1), 134–161. https://doi.org/10.1037/mgr0000004

Lee, S. S., & McKerrow, K. (2005). Advancing social justice: Women's work. *Advancing Women in Leadership, 19*, 1–2.

Lenhart, A. (2015). Teens, social media and technology overview 2015: Smartphones facilitate shifts in communication landscape for teens. *Pew Research Center*, (April), 1–47. https://doi.org/10.1016/j.chb.2015.08.026

Marshall, C., & Oliva, M. (2006). *Leadership for social justice: Making revolutions in education.* Boston, MA: Pearson Education.

Marshall, C., & Ward, M. (2004). Strategic policy for social justice training for leadership. *Journal of School Leadership, 14*(5), 530–563.

Martin, C. (2016). A library's digital Equity. *Young Adult Library Services*, (Summer), 34–37.

Marwell, E. (2017). *2017 State of the states: Fulfilling our promise to America's students.* San Francisco, CA: Education SuperHighway. Retrieved from http://stateofthestates.educationsuperhighway.org

Miranda, H., & Russell, M. (2012). Understanding factors associated with teacher-directed student use of technology in elementary classrooms: A structural equation modeling approach. *British Journal of Educational Technology, 43*(4), 652–666.

Normore, A. H., & Blanco, R. I. (2006). Leadership for social justice and morality: Collaborative partnerships, school-linked services and the plight of the poor. *International Electronic Journal for Leadership in Learning, 10*, 1–29.

Normore, A. H., & Brooks, J. S. (2014). *Educational leadership for social justice: Views from the social sciences.* Chapel Hill, North Carolina. Information Age Publishing.

NTIA. (2011). Exploring the digital nation: Computer and internet use at home. *National Telecommunications and Information Administration*, 1–56.

O'Brien, R., & Williams, M. (2016). *Global political economy: Evolution and dynamics.* New York, NY: Palgrave Macmillan.

Philip, L., Cottrill, C., Farrington, J., Williams, F., & Ashmore, F. (2017). The digital divide: Patterns, policy and scenarios for connecting the "final few" in rural communities across Great Britain. *Journal of Rural Studies, 54*, 386–398. https://doi.org/10.1016/j.jrurstud.2016.12.002

Piacciano, A. (2007). *Educational leadership and planning for technology* (4th ed.). Upper Saddle River, NJ: Pearson Education.

Rainie, L. (2016). *Digital divides 2016*. Washington, DC: Pew Research Center. https://doi.org/10.1080/15391523.2015.1080585

Rawls, J. (2001). *Justice as fairness: A restatement*. Cambridge, MA: The Belnap Press of Harvard University Press.

Resta, P., & Laferrière, T. (2015). Digital equity and intercultural education. *Education and Information Technologies, 20*(4), 743–756. https://doi.org/10.1007/s10639-015-9419-z

Rideout, V., & Katz, V. S. (2016). Opportunity for all?: Technology and learning in lower-income families (p. 48). *The Joan Ganz Cooney Center*. Retrieved from http://www.dwp.gov.uk/docs/strategyandindicators-fullreport.pdf

Ritzhaupt, A. D., Dawson, K., & Cavanaugh, C. (2012). An investigation of factors influencing student use of technology in K–12 classrooms using path analysis. *Journal of Educational Computing Research, 46*(3), 229–254. https://doi.org/10.2190/EC.46.3.b

Ritzhaupt, A. D., Liu, F., Dawson, K., & Barron, A. E. (2013). Differences in student information and communication technology literacy based on socio-economic status, ethnicity, and gender: Evidence of a digital divide in Florida Schools. *Journal of Research on Technology in Education, 45*(4), 291–307. https://doi.org/10.1080/15391523.2013.10782607

Rogers, J., Morrell, E., & Enyedy, N. (2007). Studying the struggle contexts for learning and identity development for urban youth. *American Behavioral Scientist, 51*(3), 419–443. https://doi.org/10.1177/0002764207306069

Rogers, S. E. (2016). Bridging the 21st century digital divide. *TechTrends, 60*(3), 197–199. https://doi.org/10.1007/s11528-016-0057-0

Rowsell, J., Morrell, E., & Alvermann, D. E. (2017). Confronting the digital divide: Debunking brave new world discourses how does this condition translate in schools? *The Reading Teacher, 71*(2), 157–165. https://doi.org/10.1002/trtr.1603

Scanlan, M. (2013). A learning architecture: How school leaders can design for learning social justice. *Educational Administration Quarterly, 49*(2), 348–391.

Selwyn, N., Gorard, S., & Williams, S. (2001). Digital divide or digital opportunity? The role of technology in overcoming social exclusion in U.S. education. *Educational Policy, 15*, 258–277.

Shapiro, J. P., & Stefkovich, J. A. (2016). *Ethical leadership and decision making in education: Applying theoretical perspectives to complex dilemmas* (4th ed.). New York, NY: Routledge.

Smith-Maddox, R., & Solórzano, D. G. (2002). Using critical race theory, Paulo Freire's problem-posing method, and case study research to confront race and racism in education. *Qualitative Inquiry, 8*(1), 66–84. https://doi.org/10.1177/107780040200800105

Tansley, D. (2006). Mind the gap: 2006 will witness the deepening of the digital divide. *The Financial Times, 13*, 21.

Theoharis, G. (2007). Social justice educational leaders and resistance: Toward a theory of social justice leadership. *Educational Administration Quarterly, 43*(2), 221–258. https://doi.org/10.1177/0013161X06293717

Theoharis, G. (2010). Disrupting injustice : Principals narrate the strategies they use to improve their schools and advance social justice. *Teachers College Record, 112*(1), 331–373.

Tondeur, J., Aesaert, K., Pynoo, B., Van Braak, J., Fraeyman, N., & Erstad, O. (2017). Developing a validated instrument to measure preservice teachers' ICT competencies: Meeting the demands of the 21st century. *British Journal of Educational Technology*, *48*, 462–472.

U.S. Bureau of Labor. (2016). *Office of Occupational Statistics and Employment Projections.* Washington, DC: Bureau of Labor Statistics.

U.S. Census Bureau. (2013, May). *Computer and internet use in the United States* (Report No. P20-569). Washington, DC: U.S. Government Printing Office.

van Dijk, J. A. G. M. (2006). Digital divide research, achievements and shortcomings. *Poetics*, *34*(4–5), 221–235. https://doi.org/10.1016/j.poetic.2006.05.004

Vigdor, J. L., & Ladd, H. F. (2010). Scaling the digital divide. *Working Paper 48*, (june), Online computing magazine. Retrieved from http://scholar.google.com/scholar?hl=en&btnG=Search&q=intitle:Scaling+the+Digital+Divide#1

Wang, S. K., Hsu, H. Y., Campbell, T., Coster, D. C., & Longhurst, M. (2014). An investigation of middle school science teachers and students use of technology inside and outside of classrooms: considering whether digital natives are more technology savvy than their teachers. *Educational Technology Research and Development*, *62*(6), 637–662. https://doi.org/10.1007/s11423-014-9355-4

Welner, K. G., & Weitzman, D. Q. (2005). The soft bigotry of low expenditures. *Equity and Excellence in Education*, *38*(3), 242–248. https://doi.org/10.1080/10665680591002614

CHAPTER 2

AN EXAMINATION OF THE DIGITAL DIVIDE AND ITS DIVIDING FACTORS IN FORMAL EDUCATIONAL SETTINGS

Albert D. Ritzhaupt and Tina N. Hohlfeld

In the mid-1990s, the term Digital Divide was coined to characterize a growing social inequity between those individuals who had access to Information and Communication Technology (ICT) resources and those individuals who did not. Over the past 20-years, the notion of the Digital Divide has expanded beyond mere access to ICT resources to include the knowledge, skills, and dispositions of individuals to use ICT resources to improve their quality of life. The purpose of this chapter is to provide a working definition of the Digital Divide, a brief history, a model to examine the Digital Divide in the context of formal educational settings, and a lens to focus the research that examines the "dividing factors" of the Digital Divide. Specifically, this chapter will examine the factors used to characterize the Digital Divide in formal educational settings: Socio-Economic Status (e.g., high income versus low income), education level (e.g., parents' educational level), gender (e.g., male versus female), age (generation X versus baby boomers), geography (e.g., rural versus urban), and race/ethnicity (e.g., white/Caucasian versus Hispanic). The chapter concludes with recommendations for policy-makers, educators, educational researchers, and educational administrators to address this social inequity in our society.

Crossing the Bridge of the Digital Divide: A Walk with Global Leaders, pages 19–36.
Copyright © 2019 by Information Age Publishing

INTRODUCTION

The Digital Divide remains an important problem in the 21st century within the United States. The Digital Divide traditionally describes a social-inequity between those individuals with and without access to Information and Communication Technology (ICT) resources. However, the Digital Divide has evolved over its 20-year history to incorporate the knowledge, skills, and dispositions of those individuals to use ICT resources. That is, in the United States the Digital Divide is not a question of mere access to ICT resources anymore, but rather whether individuals can meaningfully use ICT resources for the betterment of their lives and to participate in a digital democracy (Hohlfeld, Ritzhaupt, Barron, & Kemker, 2008). The Digital Divide is a multilayered phenomenon, and continues to evolve as ICT rapidly changes. While access to ICT resources remains an issue in parts of the developing world, we focus our attention in this chapter on the U.S. context.

In the 21st century, those individuals with ICT knowledge, skills, and positive attitudes (also sometimes referred to as ICT literacy) are at a distinct advantage in terms of learning in increasingly digital learning environments (NETP, 2017), competing in an increasingly digital job market (Koenig, 2011), and participating in an increasingly digital democracy (Jenkins, 2006; P21, 2017). ICT literacy is now a requirement for successful participation in virtually every aspect of our lives, including entertainment, education, work, and citizenship. While individuals need access to ICT resources, the decreasing costs of basic personal computers and laptops, and the pervasiveness of mobile devices, including smart phones and tablets, creates the ubiquitous technology environment in many schools and households in the U.S. For instance, *National Center for Education Statistics* provides the ratios of students-to-instructional computers with Internet access in formal educational settings (NCES, 2017). From 2000 to 2008, the students-to-instructional computers with Internet access ratios in all public schools decreased from 6.6 to 3.1 (NCES, 2017). The issue is now whether the individuals (e.g., students and teachers) can use these resources to improve their personal and professional lives.

As formal educational institutions (e.g., schools, universities) are often perceived as the mechanism to correct social inequities in our society, formal educational institutions have been provided with additional resources (e.g., E-Rate program) to bridge the Digital Divide. Over the years in the U.S., more teachers and students have been utilizing these resources in formal learning environments to engage in the 21st century digital activities. Nevertheless, this form of use is not always equal across the dividing factors. Although we have observed an overall decrease in the national ratio of students-to-instructional computers with Internet access in schools, we also have evidence that the computer devices may not have equitable software available for student and teacher use (Hohlfeld et al., 2008; Hohlfeld, Ritzhaupt, Dawson, & Wilson, 2017). Further, the evidence shows the way in which the ICT is used by teachers and students also varies (Hohlfeld et al., 2008; Hohlfeld et al., 2017). Research also demonstrates the knowledge, skill, and attitudes towards ICT among students (Ritzhaupt, Liu, Dawson, & Barron; 2013) and teachers (Chen & Price, 2006; Guo, Dobson, & Petrina, 2008) are different based on dividing factors.

In many ways, the Digital Divide represents a new "gap" in our educational system that is potentially more dangerous to our students, teachers, parents, administrators, legislators, and society at large. The Digital Divide is especially disconcerting, because the world is becoming more dependent upon ICT in all aspects of our lives, and consequently over time, ICT itself may inadvertently widen the achievement "gaps" (Tawfik, Reeves, & Stich, 2016). Thus, further discourse and research examining the Digital Divide in all of its forms in formal educational settings is necessary to ensure we are using ICT to narrow, as opposed to widen, the ICT "gap" between the "haves" and "have nots" in the United States. To that end, this chapter provides a brief history of the Digital Divide within the United States, a formal conceptual framework to characterize the Digital Divide within formal educational settings, a close inspection of the contemporary research on the dividing factors of the Digital Divide, and some potential solutions in the educational system to ensure we address this important problem.

BRIEF HISTORY OF THE DIGITAL DIVIDE

Since 1995, the United States (U.S.) Commerce Department's National Telecommunications and Information Administration (NTIA) published a series of reports titled *Falling through the Net* (1995, 1999, 2000). These reports analyzed computer and online access penetration rates throughout the U.S. and showed a number of dividing factors like education (e.g., high school versus bachelor), location (e.g., rural versus urban), age (e.g., young versus old), or income (e.g., rich versus poor) (NTIA, 1995). By the 1999 report, *Falling Through the Net: Defining the Digital Divide* showed soaring access rates to personal computers and the Internet in the U.S. (NTIA, 1999). However, on many characteristics (dividing factors), the NTIA found that there was still a significant, and in some cases widening Digital Divide separating ICT "haves" and "have nots" (NTIA, 1999). The original term—Digital Divide—referred to the social inequity between those who had access to computer devices and the Internet and those who did not. By the early 2000s, the term Digital Divide had become a common slogan within the education domain among policy-makers, organizations, and educators in the U.S. (Singleton & Mast, 2000).

The rise of the ICT in the U.S. resulted in the Internet economy with everyone trying to get connected, and quickly became a deeper and long-lasting phenomenon (Warschauer, 2004). Although the Federal Communications Commission (FCC) strongly supported the availability of broadband access, computer access, and training and technical assistance to as many households as possible (Barton, 2016), the dividing factors have not only persisted in the U.S., but also they have evolved. Within developed countries like the U.S., unequal distribution of ICT resulted not only in income inequality, but also unequal "education, political participation, community affairs, cultural production, entertainment, and personal interaction" (Warschauer, 2004, p. 28). The Digital Divide remains an important issue in the U.S. in the 21st century.

DIGITAL DIVIDE IN EDUCATIONAL SETTINGS

The words Digital Divide are polysemous in that they hold different meanings for different people (Ritzhaupt et al., 2013). Parents, students, educators, administrators, legislators, and librarians account differently about how they have experienced or observed the Digital Divide in their personal and professional lives (Sparks, 2013) and how it has impacted them. For this chapter, we are focusing on two primary stakeholders within the educational system: students and teachers. While the literature provides many definitions of the Digital Divide, in this chapter, we provide the following operational definition for formal educational settings: The Digital Divide is a social inequity due to disparate quantity and/or quality of teachers' and students' access, use, and creation of original artifacts with Information and Communication Technology (ICT) resources. With this definition in mind, there are some important terms that emphasize the perspective presented in this chapter.

First, social inequity refers to unequal opportunities for engagement in society (e.g., social, economic, political, educational, or personal pursuits) based on different statuses or groups (e.g., location, socio-economic status, race/ethnicity, age, disability, or education level). Second, the use of the words "access", "use", and "creation" are deliberately linked to the conceptual model (*Levels of the Digital Divide in Schools*). Third, the quality and quantity of students' interactions with ICT bring about the multi-layered phenomenon of the Digital Divide with each layer associated with different problems, research methods, and solutions. Fourth, ICT resources include both physical (e.g., computer, tablet, smart phone) and digital (e.g., software, applications, media, and Internet) resources that can be utilized to create original artifacts. Finally, by "original artifacts" we refer to the many types of objects that can be created by students and teachers with ICT, including original artwork, lesson plans, digital music, written publications, opensource software, animations, videos, games, blogs, web pages, presentations, spreadsheets, and much more.

Firmly grounded in the definition provided, we use the *Levels of the Digital Divide in Schools* presented by Hohlfeld et al. (2008). Figure 1 provides a modified visualization of the conceptual model of the Digital Divide for formal educational settings. The model provides three layers, starting with school infrastructure and access to ICT, moving to the classroom with teacher and student use of ICT, and finally, presenting the individual empowerment of the teachers and students using ICT as the highest layer. Activities, research, problems, and solutions vary at each of these levels. From our operational definition, we use the terms "access" at layer one, "use" at layer two, and "creation" at layer three. The underlying assumption

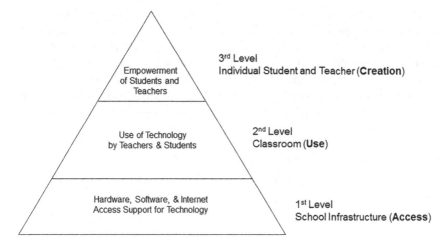

FIGURE 2.1. Modified Levels of the Digital Divide in Schools (Hohlfeld et al., 2008).

of this model is that student and teacher creation of relevant artifacts using ICT and their ultimate empowerment with ICT is a desirable outcome for society.

Level One: School Infrastructure and Access

Access to appropriate levels of ICT resources in formal educational settings requires extensive planning and funding from educational administrators and other relevant stakeholders with a vested interest (Ritzhaupt, Hohlfeld, Barron, & Kemker, 2008). Level one of the conceptual model includes providing access to appropriate ICT resources within a school, including hardware, software, Internet access, and support structures for both teachers and students. School infrastructure and access is intentionally layered at the bottom of the model to reflect that access to ICT resources is an antecedent to teacher and student use, and ultimately, their empowerment with ICT. Early research on the Digital Divide often emphasized counting the number of instructional computers available to students with Internet access (e.g., Hess & Leal, 2001; Valadez & Duran, 2007). This metric was used as a sign of access to ICT within formal educational settings.

The Digital Divide manifests itself at the first level of school infrastructure and access by limiting the quantity and quality of opportunities to use ICT resources either in the school setting, or in the teachers' or students' homes. In particular, students from lower-income homes, rural homes, ethnically diverse homes, and homes with parents with lower levels of educational attainment are less likely to have broadband Internet access (NCES, 2017). In formal school settings, educational institutions have dramatically improved access to ICT resources for students and teachers during the school day (Hohlfeld et al., 2008; Hohlfeld et al., 2017). However, providing equitable access to Internet-enabled devices at school does not guarantee that these ICT resources will be used equitably by students and teachers (Cuban, 2009). Further, there are no guarantees that students will have access to ICT resources in their homes to complete digital homework assignments or online enhancement activities, unless the school provides ICT resources via special initiatives like provision of an assigned one-to-one device for use in both school and home.

Level Two: Classroom Use by Teachers and Students

While access to ICT is still a problem in formal educational settings, educators, administrators, and educational researchers have become more concerned with how these ICT resources are used in the classroom setting. Empirical research evidence demonstrates that students and teachers in lower-income schools are using technology for different purposes. For example, students in low-income schools are found to use technology more for computer-directed activities such as individualized tutoring, remedial academic skills attainment, or drill-and-practice proficiency reinforcement, while their higher-income counterparts are using ICT more for student-controlled activities such as creating original artifacts and communicating knowledge and ideas with ICT (see Hohlfeld et al., 2008; Hohlfeld et al., 2017). These findings show evidence of a Digital Divide manifesting at the second level of the model which focuses on both teacher and student use of ICT in educational settings.

The second level of the Digital Divide presents different types of complications and research applications. For instance, Cuban, Kirkpatrick, and Peck (2001) examined how the ICT resources were actually being used by students and teachers in their classroom environment across dividing factors (e.g., rural versus urban or High-SES versus Low-SES). While legislators and administrators might invest heavily to integrate the ICT resources into schools and classrooms, if the teachers are not trained to meaningfully integrate ICT into the students' instruction (e.g., sufficient professional development), do not have access to adequate technology support (e.g., technology specialist in a school), and do not support the mission of the ICT program (e.g., leadership), the Digital Divide may manifest as inequitable learning experiences with the ICT resources for the students. These essential conditions are outlined by the International Society for Technology in Education (ISTE) and are perceived as necessary elements to effectively leverage ICT for teaching and learning (ISTE, 2017).

Level Three: Individual Creation by Students and Teachers

The highest level of the conceptual model requires that both teachers and students have the knowledge, skills, intent, and dispositions to create original artifacts with ICT resources. This perspective emphasizes productive and creative uses of ICT by students and teachers as opposed to consumption and reactive uses of ICT, such as using ICT for online standardized testing and remedial skills programs. The ultimate goal of meaningfully integrating ICT resources into schools (Level 1) and classrooms (Level 2) is to empower teachers and students to participate in an increasingly digital society. Empowered teachers can support their students in achieving their personal and professional goals. However, educational administrators must first nurture their teachers' positive beliefs about the benefits of ICT and support and promote their enhanced ICT integration practices (Ertmer, Ottenbreit-Leftwich, Sadik, Sendurur, & Sendurur, 2012). ICT has the potential to support, advance, and enrich opportunities and outcomes for all students. ICT literacy and the ability to leverage ICT for learning are essential to the future empowerment of all students across dividing factors. In the next section, we review some research on the dividing factors of the Digital Divide as it relates to formal educational settings.

DIVIDING FACTORS RELATING TO THE DIGITAL DIVIDE

The dividing factors refer to the different statuses or groups (e.g., geography, socio-economic status, race/ethnicity, gender, age, or education level) that have been associated with differences in ICT access, use, and literacy, and form the basis of the Digital Divide. This section reviews seven of the dividing factors that have been documented in educational research. While this is not an exhaustive list and we acknowledge other dividing factors exist (e.g., disability status, culture, or English language learners), we provide this overview to highlight factors that are commonly connected with social inequity. Because ICT access, use, and literacy have been linked to important outcomes, such as per capita income (Pew, 2015), understanding the current research literature in formal educational settings is imperative to advance the discourse concerning the Digital Divide.

Socio-Economic Status

Socio-Economic Status (SES) refers to the income levels of the student, teacher, family (e.g., parents or caregivers), school, and community. Income is an understandable dividing factor in that if individuals, families and communities do not have the financial resources to spend on ICT, they will likely have less ICT access, use, and literacy. Digital Divide research which examines equity by SES status have used the income levels of individual students and teachers, the overall family income, the SES levels of a school (e.g., percentage of students on free-and-reduced price lunch), and the overall income levels of the communities in which the students and teacher reside. The income of a family, as of 2013, still moderates ICT access. Specifically, 72% percent of people with family incomes from $40,000 to $49,999 used the Internet, compared to 85 percent of people with family incomes of $100,000 or more (NCES, 2017). When using the school level SES as the unit of analysis, empirical evidence show differences between both teacher and student use of ICT resources within schools (Hohlfeld et al., 2008; Hohlfeld et al., 2017). More specifically, teachers in lower-income schools were less likely to use software for instructional purposes, and students were less likely to use computers for student-directed activities like creating authentic resources and communicating with ICT (Hohlfeld et al., 2017). When looking at the student as the unit of analysis, higher-income middle school students significantly outperformed their lower-income counterparts on a performance based ICT literacy assessment grounded on the ISTE standards (Ritzhaupt et al., 2013).

Educational Level

Educational level refers to the educational attainment of the teacher, student, parents, and community in which they reside. Often in educational research, educational researchers will use the parents' educational attainment as an independent or moderating variable to examine outcomes. For instance, in the 2014 administration of the *Technology and Engineering Literacy* (TEL) assessment, 55% of the students with parent's that graduated college scored at or above proficient, while only 20% of the students with parent's that did not finish high school scored at or above proficient (NRC, 2014), showing a rather large gap between the two groups. At an individual level, 54% of persons who had not completed high school used the Internet, compared with 64% of those who had completed only a high school diploma or equivalent, and 89% of those with a bachelor's or higher degree (NCES, 2017). Reynolds and Chiu (2016) showed that middle and high school student home computer use was statistically influenced by a parent's education level.

Gender

Gender refers to the state of being male or female, and generally refers to the biological sex of the individual. Much empirical research has investigated the influence of gender on ICT literacy and attitudes of both students and teachers. Earlier accounts of research on gender and ICT literacy and attitudes have shown that males have higher scores than their female counterparts, particularly on self-report measures (Murphy, Coover, & Owen, 1989; Wilder, Mackie, & Cooper, 1985). These gender differences have led to stereotypes and misunderstandings among the educational community about gender-roles. However, in more recent literature, females have outperformed their male counterparts on ICT literacy performance assessments (Hohlfeld, Ritzhaupt, & Barron, 2013; Ritzhaupt et al., 2013). Further, when controlling for attitudinal factors, the differences between genders was no longer significant in one large-scale study (Hohlfeld et al., 2013). In fact, the most recent scores from the 2014 administration of the TEL assessment showed females scored higher than their male counterparts (NRC, 2014). However, the number from the computing workforce tells a different story with only 26% of women serving in professional computing roles and 18% of women earning computer science bachelor degrees (NCWIT, 2017).

Age

Age was stated as a dividing factor early in the conception of the Digital Divide with younger generations (e.g., Millennials) typically being compared with older generations (e.g., Baby Boomers) with the assumption that the older generation would have less access, use, and ICT literacy. One must be careful in examining the influence of age, especially in light of the Digital Native hypothesis (Prensky, 2001), which has largely been debunked in the empirical research literature on the topic (Kirschner & van Merriënboer, 2013). In addition, Reeves and Oh (2008) showed that many of the survey tools used in the research about age differences lacked sufficient validity and reliability evidence. Nevertheless, age does appear to moderate access, use, and ICT literacy for both teachers and students in several empirical studies on the topic. For instance, there appears to be a negative correlation between the number of years of teaching experience (or age) and key technology measures, including teacher use of technology (Liu, Ritzhaupt, Dawson, & Barron, 2017; Ritzhaupt, Dawson, & Cavanaugh, 2012), teacher confidence and comfort using technology (Liu et al., 2017), computer proficiency (Inan, & Lowther, 2010), and classroom technology integration (Liu et al., 2017). Reynolds and Chiu (2016) showed that age moderated basic computer activities scores of middle and high school students in West Virginia. Oh and Reeves (2014) provide a comprehensive review of literature on the differences between generations. They reported that researchers agree that students use ICT resources differently in school, where they use technology for more basic tasks, than they do out of school. They suggest that the differences in how technology and ICT resources are used at schools is not based on age, but rather other dividing factors. Oh and Reeves (2014) conclude that there is a need for more generational differences, or age, ICT research across different groups.

Geography

The physical location of a teacher and student can moderate their access, use, and ICT literacy, depending on the ICT infrastructure and population density of the location. Put simply, zip codes still matter in the U.S. Whitacre and Mills (2007) demonstrated that the high-speed Internet access is moderated by geography in the early 2000s in the U.S. with differential rates between rural and urban locations favoring the urban centers. More recently, Goh and Kale (2016) examined the rural versus urban gap in West Virginia schools on Web 2.0 access and skill measures, and found that both physical access and usage access significantly favored the urban school teachers over the rural school counterparts. However, when taking into account the recent administration of the TEL assessment, 45% of the student participants in rural regions scored at proficient whereas only 37%

in the cities scored at proficient (NRC, 2014). Access to broad-band Internet connections appears to still be an issue in rural communities. Other dividing factors (e.g., SES) may be interacting with geography to moderate these results.

Race/Ethnicity

Race/ethnicity has been linked to ICT access, use, and literacy measures within the U.S. Generally speaking, there are gaps between White/Caucasian students and their non-White counterparts. For instance, Reynolds and Chiu (2016) showed statistically significant differences in school computer use among Hispanic/Latino middle and high school students in West Virginia. When examining Internet use among White, Black, and Hispanic populations age three and over, the percentage of Internet users was highest among Whites (75%), followed by Blacks (64%), and then Hispanics (61%) (NCES, 2017). Campos-Castillo (2015) provides evidence from a large sample in the U.S. in 2012 that White participants have the highest level of Internet access, followed by Latino participants, and then Black participants. Ritzhaupt et al. (2013) showed a wide performance difference between a sample of more than 5,000 White and non-White middle school students from Florida on an ICT literacy performance assessment favoring the White students in the sample. On the recent administration of the TEL assessment, 56% of White students scored at proficient, 28% of Hispanic students scored at proficient, and 18% of Black students scored at proficient.

Summary of Dividing Factors

This section has provided a sample of the dividing factors commonly used to characterize the Digital Divide in the U.S. While not all factors were discussed (e.g., disability status, culture, or English language learner), the factors that were highlighted demonstrate that the Digital Divide empirical research manifests itself differently at varying levels on each of the dividing factors noted. As our efforts as a society can both narrow or widen the Digital Divide, it becomes tricky to identify the independent versus dependent variables in the research models. Is the technology outcome measure the independent or dependent variable? Or is the dividing factor still significant while controlling for other factors? The empirical research can go in both directions. However, one notable observation is that future research on the Digital Divide should attempt to simultaneously control for as many of the dividing factors as possible in the research model. There are potential interaction effects associated with the dividing factors that presently remain unexplained in much of the empirical research. For instance, geography as a dividing factor may show as insignificant in the model if controlling for other variables like SES. More empirical research is necessary to address this issue.

BRIDGING THE DIGITAL DIVIDE IN
FORMAL EDUCATIONAL SETTINGS

Narrowing the Digital Divide in formal education settings involves assuring that the "Essential Conditions" stated by the International Society for Technology Education (ISTE) are instituted for all teachers and students, irrespective of the dividing factors. Just as there are levels of the Digital Divide, potential solutions addressing the Digital Divide requires coordination of activities at three levels. Educators, policy-makers, educational researchers, and educational administrators can focus on bridging the Digital Divide with a systems-view of programs at the mega (e.g., international or national), macro (e.g., state, municipal, or school district), and micro (e.g., schools and classrooms) levels.

Mega Level Solutions: International and National Leadership Role

Leadership, vision, and direction often begins at the mega level. For example, ISTE, an international professional organization has developed technology standards for administrators, teachers, coaches, and students, which have been widely adopted by policy-makers and administrators across the U.S. Further, the U.S. *Department of Education's Office of Educational Technology* publishes the *National Educational Technology Plan* (NETP), which outlines an overall strategy for ICT resources in educational settings. In turn, departments of the U.S. government financially support building the ICT infrastructure and developing model programs which address the Digital Divide in schools by partnering with universities, and providing competitive grants or ICT resources. For instance, each year the National Science Foundation (NSF) offers the competitive Innovative Technology Experiences for Students and Teachers (ITEST) grants (NSF, 2017).

Government ICT programs (e.g., E-Rate program) at the mega level can provide discounts for telecommunications and Internet access costs for schools and libraries to ensure equitable access across the dividing factors. Although some evaluations of the E-Rate program within the U.S. concluded that the program had failed to close the Digital Divide (Park, Sinha, & Chong, 2007), by Fall of 2001, 99% of public schools in the U.S. had access to the Internet (NCES, 2017). Many formal educational institutions are now concentrating on providing wireless Internet access throughout the facility for students and teachers to have ubiquitous and uninterrupted access to Internet resources.

At the mega level, national corporations also have supported programs to address the several levels of the Digital Divide in formal education. For instance at first level of the Digital Divide (infrastructure), Google has partnered with the U.S. government to provide Google Fiber, high-speed Internet access to low-income families, allowing children to continue school activities at home (Newcomb, 2015). At the next level (classroom), the Khan Academy, provides free Open Educational Resources (OER) online for teachers to enhance their curriculum with high-quality learning experiences for their students. To address Level two, national and international non-profit professional organizations, such as ISTE, offer training and professional development resources to support teachers delivering high-quality digital learning experiences.

Key, at the mega level, is supporting large scale programs, research projects, and evaluation systems to investigate disparities and changes in the Digital Divide by the many dividing factors and disseminating the results. For instance, *The Program for International Student Assessment* (PISA) sponsored by the Organization for Economic Co-operation and Development (OECD) includes ICT components in the assessments used with students, and the *National Assessment of Educational Progress* (NAEP) by *National Center for Education Statistics* (NCES) includes the *Technology and Engineering Literacy* (TEL) assessment. In addition, this educational data has been archived for future Digital Divide research by making it publically accessible to educational researchers for secondary data mining, analysis, discovery, and dissemination.

Macro Level Solution: Leadership at the State and District Level

In the U.S., educational policy is determined at the macro level of state governments. States should create an ICT plan to set the vision for the schools and school districts in the state. Following the leadership provided at the mega level, the state legislature sets the educational standards for student outcomes and requirements for teacher certification. State governments set the course requirements, specify the curriculum for pre-service teacher education in collaboration with institutions of higher education, and administer certification assessments to assess the content knowledge of the teachers. States also set the continuing education requirements for teachers to maintain their professional certification. As a result, states have a major impact on the curriculum and teacher preparedness, which addresses the second and third levels of the Digital Divide. Recent research showed that less than 50% of the U.S. state departments of education offered educational technology certifications for teachers (Ritzhaupt, Levene, & Dawson, 2017).

Together, the state governments with municipal governments raise the revenue to accomplish positive educational outcomes for all students and teachers. State and municipal governments can earmark specific revenue for special ICT programs, which are designed to overcome the Digital Divide in local communities. These programs may address all three levels of the Digital Divide. For instance, funds can be allocated for technology magnet school programs, special professional development programs for teachers which address technology integration, or technology-enhanced student assessment systems.

School districts adopt local policies for program implementation at the macro level. ICT policies can facilitate the flow of instruction between school and home. For instance, school districts can adopt BYOD ("Bring Your Own Device") programs (Raths, 2012) where students bring their own device to connect to the school's network while at school and then use them at home. When school districts are concerned that the students do not own their own devices, they can implement one-to-one device programs where each student in a school is provided a computing device to engage in digital teaching and learning at school and home (Dawson, Cavanaugh, & Ritzhaupt, 2008). Many school districts support the flow of instruction between schools and homes by providing cloud-based software solutions (e.g., Google Drive or Microsoft 360), which give teachers and students consistent access to the same relevant instructional software at school and home.

As noted, it is not merely access to ICT per se that narrows Digital Divide but how ICT resources are used by students and teachers. We have years of evidence that shows merely placing ICT resources in schools does not lead to meaningful changes in important teacher or student outcomes (Cuban, 1986, 2009). To address the second level of Digital Divide, it is imperative to provide rich job-embedded professional development opportunities for teachers which help them develop their ICT skills and improve their ICT integration knowledge, skills, and dispositions (Ritzhaupt et al., 2013). School districts are charged with providing meaningful professional development experiences for teachers; thus, school districts must engage teachers in relevant, sustained, active, collaborative, and content-focused professional development opportunities (Desimone, 2009). Empowering teachers is the key to unlocking our students' potential for effective ICT integration (Wenglinsky, 2005).

Micro Level Solutions: Leadership and Implementation in the School

The schools, largely influenced by the principal, and the classrooms, largely influenced by the teacher, implement the policy mandated by mega and macro authorities, deliver the educational curriculum, and assess student outcomes. The schools have the ultimate responsibility for maintaining the ICT infrastructure and directing the ICT resources efficiently and effectively to their specific programs. The ISTE standards come to the forefront and call upon administrators to set the vision, allocate the appropriate resources, and assess the outcomes of ICT programs in their schools. Important to note is that equitable use of ICT does not necessarily mean equal use. Principals must convey visionary leadership and establish a digital age learning culture within the school setting which empowers the teachers at the third level of the Digital Divide (ISTE, 2017). Equitable use involves the teachers choosing how and when to support their students with the appropriate resources that enhance their students' ICT literacy at the second and third levels of the Digital Divide. Teachers will need to serve as facilitators and collaborators both with their students and other teachers, and ultimately, design, implement, and model meaningful and relevant ICT experiences for their students to thrive (ISTE, 2017). Students also have an important role to play to ensure their success. They must be active in the learning process by setting personal learning goals, utilizing ICT as a good citizen, and thinking creatively when using ICT resource to create authentic artifacts.

Schools, just as the mega and macro levels, should invest in developing an ICT plan aligned with their mission and vision as a school. All stakeholders should be involved in the ICT planning process, including students, teachers, administrators, parents, local community members, school technology personnel, and local business leaders. This ICT plan should be revisited annually by the stakeholders to ensure goals, objectives, and assessment outcomes are aligned. Administrators must acknowledge that ICT planning and funding is not a one-time expense, but rather an ongoing investment into the schools vitality (Ritzhaupt et al., 2008). Both teachers and administrators must work to engage parents in the education of their children, by creatively enhancing family communication with ICT along with other traditional and familiar communication modes (Hohlfeld, Ritzhaupt, & Barron, 2010).

School Leadership to Bridge the Digital Divide

Bridging the Digital Divide in the 21st century requires visionary school administrators to think beyond the students and teachers in their schools to their local communities and society at large. ISTE provides administrator standards that include expectations in the areas of visionary leadership, digital age learning culture, excellence in professional practice, systemic improvement, and digital citizenship. Put simply, educational administrators have immense pressure and expectations from their local communities and society to assist with creating 21st century citizens and close the Digital Divide. The key to influencing and implementing any social change is the personal contact of an individual with others. At the ground level for implementing changes in the Digital Divide is the personal influence of the school administrator during individual interactions with students, teachers, other administrators, the school board directors, the local community leaders, and contacts in professional organizations.

First, educational administrators, like teachers, need to engage in professional development opportunities to enhance their knowledge, skills, and attitudes. They apply this knowledge and skill to ensure a robust infrastructure at Level One of the Digital Divide. This involves careful planning, and creatively securing and appropriately allocating ICT resources by working with a diverse set of stakeholders (Ritzhaupt et. al., 2008). Educational administrators further support the ICT professional development opportunities for their teachers to encourage excellence in professional practice in the classroom. They also model the appropriate behaviors and uses of ICT to both students and teachers within their school setting. These aforementioned activities assist to address Level Two of the Digital Divide.

Educational administrators reach out to their local communities to engage public (e.g., school boards and municipal governments), private (e.g., local business and industry), and non-profit (e.g., professional organizations and foundations) for support in their school's mission for narrowing the gap. They share their school resources with the local community and parents by providing ICT awareness and training, supporting initiatives that support families by allowing students to bring devices home or to school, and supporting ongoing communication with the families and the community (Hohlfeld et al., 2010). They share these activities with their teachers and students to model behaviors that support engagement of students and teachers at Level Three of the Digital Divide. They share these activities with their professional networks to provide examples of visionary leadership aimed at closing the Digital Divide to encourage the efforts of others, which makes their contributions global in impact.

CLOSING REMARKS

The Digital Divide remains an important problem for the educational system in the U.S. Though the construct has evolved from an issue of ICT access to one of use, knowledge, skills, and dispositions, participating in the 21st century, overcoming the Digital Divide in schools requires both teachers and students to have proficient levels of ICT literacy. E-Commerce, e-learning, and e-government are now a reality in the U.S. We must ensure that we are not leaving groups behind based on the dividing factors (e.g., location, gender, income, age, or race). The current evidence from the literature suggests that Digital Divide is still a persistent issue for many communities in the U.S.

Because ICT has the potential to widen the gap between the "haves" and "have nots" based on the dividing factors, policy-makers, administrators, and educational researchers must carefully plan, implement, and monitor the ICT integration patterns at all levels of the Digital Divide in our formal educational settings. Education is perhaps our best solution to this social inequity. Nevertheless, we must ensure that we are not contributing to the problem, and widening the gap, or even worse, contributing to additional gaps in our educational system. Educational programs aimed at addressing the Digital Divide, should tackle all three levels of the Digital Divide model presented in this chapter. Neglecting the multilayered nature of the Digital Divide, can have serious adverse ramifications on the outcomes of the program.

ICT literacy programs and assessments should align with the ultimate goal of the creation of "original artifacts" as the target of empowerment with ICT for teachers and students. Although many individuals can consume media from ICT resources, some are not capable of using ICT resources for creation and empowerment tasks. We firmly believe that we need to embrace the producer model, in which students and teachers become the creators of authentic ICT resources. As highlighted in the new version of the ISTE standards for students, we need individuals who are knowledge constructors that can innovatively design and computationally think to solve real-world problems (ISTE, 2017).

Narrowing, and ultimately, closing the Digital Divide requires involvement at all levels of the U.S. public and private sectors. While formal education is a solution to the Digital Divide, it can also be part of the problem. Neglecting this phenomenon could have catastrophic effects on our country's vitality in the next decades. The goal of formal education is for all students and teachers to harness the power of ICT to improve their personal livelihoods and enhance society. We hope this chapter is a starting place to engage in the discourse necessary to influence positive changes in our local educational communities.

REFERENCES

Barton, J. (2016). *Closing the digital divide: A framework for meeting CRA obligations.* Federal Reserve Bank of Dallas, Retrieved from https://EconPapers.repec.org/RePEc:fip:feddmo:00004

Campos-Castillo, C. (2015). Revisiting the first-level digital divide in the United States: Gender and race/ethnicity patterns, 2007–2012. *Social Science Computer Review, 33*(4), 423–439.

Chen, J. Q., & Price, V. (2006). Narrowing the digital divide: Head start teachers develop proficiency in computer technology. *Education and Urban Society, 38*(4), 398–405.

Cuban, L. (1986). *Teachers and machines: The classroom use of technology since 1920.* New York, NY: Teachers College Press.

Cuban, L. (2009). *Oversold and underused.* Boston, MA: Harvard University Press.

Cuban, L., Kirkpatrick, H., & Peck, C. (2001). High access and low use of technologies in high school classrooms: Explaining an apparent paradox. *American Educational Research Journal, 38*(4), 813–834.

Dawson, K., Cavanaugh, C., & Ritzhaupt, A. D. (2008). Florida's EETT leveraging laptops initiative and its impact on teaching practices. *Journal of Research on Technology in Education, 41*(2), 143–159.

Desimone, L. M. (2009). Improving impact studies of teachers' PD: Toward better conceptualization and measures. *Educational Researcher, 38,* 181–199.

Ertmer, P. A., Ottenbreit-Leftwich, A. T., Sadik, O., Sendurur, E., & Sendurur, P. (2012). Teacher beliefs and technology integration practices: A critical relationship. *Computers & Education, 59*(2), 423–435.

Guo, R. X., Dobson, T., & Petrina, S. (2008). Digital natives, digital immigrants: An analysis of age and ICT competency in teacher education. *Journal of Educational Computing Research, 38*(3), 235–254.

Goh, D., & Kale, U. (2016). The urban–rural gap: Project-based learning with Web 2.0 among West Virginian teachers. *Technology, Pedagogy and Education, 25*(3), 355–376.

Hess, F. M., & Leal, D. L. (2001). A shrinking "digital divide"? The provision of classroom computers across urban school systems. *Social Science Quarterly, 82*(4), 765–778.

Hohlfeld, T. N., Ritzhaupt, A. D., Barron, A. E., & Kemker, K. (2008). Examining the digital divide in K–12 public schools: Four-year trends for supporting ICT literacy in Florida. *Computers & Education, 51*(4), 1648—1663.

Hohlfeld, T. N., Ritzhaupt, A. D., & Barron, A. E. (2010). Connecting schools, community, and family with ICT: Four-year trends related to school level and SES of public schools in Florida. *Computers & Education, 55*(1), 391–405.

Hohlfeld, T. N., Ritzhaupt, A. D., & Barron, A. E. (2013). Are gender differences in perceived and demonstrated technology literacy significant? It depends on the model. *Educational Technology Research and Development, 61*(4), 63–663.

Hohlfeld, T. N., Ritzhaupt, A. D., Dawson, K., & Wilson, M. L. (2017). An examination of seven years of technology integration in Florida schools: Through the lens of the levels of digital divide in schools. *Computers & Education, 113*, 135–161.

Inan, F. A., & Lowther, D. L. (2010). Factors affecting technology integration in K–12 classrooms: A path model. *Educational Technology Research and Development, 58*(2), 137–154.

ISTE (2017). *ISTE essential conditions.* Retrieved from: https://www.iste.org/standards/essential-conditions

Jenkins, H. (2006). *Convergence culture: Where old and new media collide.* New York, NY: NYU press.

Kirschner, P. A., & van Merriënboer, J. J. (2013). Do learners really know best? Urban legends in education. *Educational Psychologist, 48*(3), 16–183.

Koenig, J. A. (Ed.). (2011). *Assessing 21st century skills: Summary of a workshop.* Washington, DC: National Research Council, National Academies Press.

Liu, F., Ritzhaupt, A. D., Dawson, K., & Barron, A. E. (2017). Explaining technology integration in K-12 classrooms: A multilevel path analysis model. *Educational Technology Research and Development, 65*(4), 795–813.

Murphy, C. A., Coover, D., & Owen, S. V. (1989). Development and validation of the computer self-efficacy scale. *Educational and Psychological Measurement, 49*(4), 893–899.

NCES (2017). *National Center for Educational Statistics.* U.S. Department of Education. Retrieved from, https://nces.ed.gov/

NCWIT (2017). *By the numbers.* National Center for Women and Information Technology. Retrieved from, https://www.ncwit.org

Newcomb, A. (2015). *Google Fiber helping to close the digital divide.* ABC News. Retrieved from, http://abcnews.go.com/Technology/google-fiber-helping-to-close-digital-divide/story?id=32492506

NETP (2017). *National educational technology plan.* Office of Educational Technology, United States Department of Education. Retrieved from, https://tech.ed.gov/netp/

NRC (2014). *Technology & engineering literacy (TEL). National report card.* Retrieved from, https://www.nationsreportcard.gov/tel_2014/

NSF (2017). *Innovative technology experiences for students and teachers (ITEST).* National Science Foundation. Retrieved from, https://www.nsf.gov/funding/pgm_summ.jsp?pims_id=5467

NTIA (1995). *Falling through the net: A survey of the" have nots" in rural and urban America.* Washington, DC: National Telecommunications and Information Administration, US Department of Commerce.

NTIA (1999). *Falling through the net: Defining the digital divide. A Report on the Telecommunications and Information Technology Gap in America.* Washington, DC: National Telecommunications and Information Administration, US Department of Commerce.

NTIA (2000). *Falling through the net: Toward digital inclusion.* Washington, DC: National Telecommunications and Information Administration, US Department of Commerce.

Oh, E., & Reeves, T. C. (2014). Generational differences and the integration of technology in learning, instruction, and performance. In *Handbook of research on educational communications and technology* (pp. 819–828). New York, NY: Springer,

P21 (2017). *Partnership for 21st Century Skills.* Retrieved from, http://www.p21.org

Park, E., Sinha, H., & Chong, J. (2007). Beyond access: An analysis of the influence of the e-rate program in bridging the digital divide in American schools. *Journal of Information Technology Education, 6,* 387–406.

Pew (2015). *Internet access strongly related to per capita income.* Pew Research Center. Retrieved from, http://www.pewglobal.org/interactives/internet-usage/

Prensky, M. (2001). Digital natives, digital immigrants part 1. *On the Horizon, 9*(5), 1–6.

Raths, D. (2012). Are you ready for BYOD: Advice from the trenches on how to prepare your wireless network for the bring-your-own-device movement. *The Journal (Technological Horizons in Education), 39*(4), 28.

Reeves, T. C., & Oh, E. (2008). Generational differences. *Handbook of research on educational communications and technology, 3,* 295–303.

Reynolds, R., & Chiu, M. M. (2016). Reducing digital divide effects through student engagement in coordinated game design, online resource use, and social computing activities in school. *Journal of the Association for Information Science and Technology, 67*(8), 1822–1835.

Ritzhaupt, A., Hohlfeld, T. N., Barron, A. E., & Kemker, K. (2008). Trends in technology planning and funding in Florida K–12 schools. *International Journal of Education Policy and Leadership, 3*(8), 1–17.

Ritzhaupt, A. D., Dawson, K., & Cavanaugh, C. (2012). An investigation of factors influencing student use of technology in K–12 classrooms using path analysis. *Journal of Educational Computing Research, 46*(3), 229–254.

Ritzhaupt, A. D., Levene, J. & Dawson, K. (2017, April). *Where are we as a certifiable body of knowledge? Technology certificates and endorsements offered by state departments of education in the United States.* Paper presented at the American Educational Research Association, San Antonio, TX.

Ritzhaupt, A. D., Liu, F., Dawson, K., & Barron, A. E. (2013). Differences in student information and communication technology literacy based on socio-economic status, ethnicity, and gender: Evidence of a digital divide in Florida schools. *Journal of Research on Technology in Education, 45*(4), 291–307.

Sparks, C. (2013). What is the "digital divide" and why is it important? *Javnost—The Public, 20*(2), 27–46.

Singleton, S., & Mast, L. (2000). How does the empty glass fill? A modern philosophy of the Digital Divide. *Educause Review, 35*(6), 2–36.

Tawfik, A. A., Reeves, T. D., & Stich, A. (2016). Intended and unintended consequences of educational technology on social inequality. *TechTrends, 60*(6), 598–605.

Valadez, J. R., & Duran, R. (2007). Redefining the digital divide: Beyond access to computers and the Internet. *The High School Journal, 90*(3), 31–44.

Warschauer, M. (2004). *Technology and social inclusion: Rethinking the digital divide.* Cambridge, MA: MIT Press.

Wenglinsky, H. (2005). Technology and achievement: The bottom line. *Educational Leadership, 63*(4), 29.

Whitacre, B. E., & Mills, B. F. (2007). Infrastructure and the rural–urban divide in high-speed residential Internet access. *International Regional Science Review, 30*(3), 24–273.

Wilder, G., Mackie, D., & Cooper, J. (1985). Gender and computers: Two surveys of computer-related attitudes. *Sex Roles, 13*(3), 215–228.

CHAPTER 3

NOT ALL YOUNG PEOPLE "USE" THE INTERNET

Exploring the Experiences of Ex-Use Amongst Young People in Britain

Rebecca Eynon and Anne Geniets

Despite the rhetoric about 'digital youth' a great deal of research has highlighted how young people vary significantly in the ways that they engage with the Internet. A rarely explored group is those young people who hardly use the Internet at all. Yet, they are an important focus of attention, particularly as many services and support for young people are often digital by default. Through in-depth interviews with 22 young people this chapter explores the experiences of young people who consider themselves "ex-users" of the Internet. Through the analysis we identify two groups of young people in this category: young people who used to be Internet users, but no longer feel they are Internet users; and those who had never really felt like Internet users in the first place. We conclude by highlighting the continuing challenges of measuring 'use' of the Internet and the potential policy and practice implications for this group of young people and for global leaders.

Crossing the Bridge of the Digital Divide: A Walk with Global Leaders, pages 37–55.

INTRODUCTION

For decades there has been a seemingly impervious rhetoric around young people and digital technology that inaccurately depicts all young people has being highly digitally enabled (for an in-depth discussion on this issue, see Davies & Eynon, 2013; Facer & Furlong, 2001; Hargittai, 2010; Helsper & Eynon,2010). Similar to the rest of the UK population young people have a variety of different levels of engagement with the Internet and use it for an array of different purposes. Research has highlighted a complex array of factors that help us to understand in what ways and to what extent people use the Internet (DiMaggo, Hargittai, Neumann & Robinson 2001; Van Dijk 2006). The lack of recognition of this varied form of engagement among young people means that perhaps less attention has been placed on this group in the digital inequality literature than for other people at different life stages.

Indeed, a dichotomous divide between users and non-users of technology—constructed +primarily by policy makers designed to measure uptake and the success of digital strategies—has given way to recognition that digital inequality is a multifaceted issue. Dichotomous measures of use and non-use have long been rejected by academics in the field, as they reduce understanding to, "a simple and singular boundary between the digitally engaged and those who are disengaged" (Halford & Savage 2010, p. 937); and in the case of young people "homogenizes the experiences of the more and less advantaged segments of what is implicitly presumed to be a uniformly 'wired' population" (Robinson, 2009, p. 490). A number of researchers have highlighted the blurred boundaries between the two seemingly dichotomous categories of use and non-use of the Internet and also highlighted how this can often change over time (Haddon, 2004; Murdock, 2002; Selwyn, 2006; Wellman & Haythornthwaite, 2002; Wyatt, Oudshoorn, & Pinch, 2003).

Studies of non-use have been an important (if relatively niche) area of study (Baumer, Ames, Burrell, Brubaker & Dourish 2015), with a number of authors bringing attention to this issue within discussions of the digital divide (e.g., Reisdorf & Groselj, 2017; Verdegem & Verhoest, 2009). Some have provided typologies of different forms of non- use. For example, Satchell and Dourish identify six forms of non-use: lagging adoption, active resistance, disenchantment, disenfranchisement, displacement (i.e. a form of proxy use), and disinterest (Satchell & Dourish 2009 p. 9). Wyatt and colleagues identified four groups: resistors, rejectors, excluded and expelled (Wyatt, Thomas, & Terranova, 2002); as did Lenhart and Horrigan: intermittent users, net dropouts, net evaders and the truly unconnected (Lenhart & Horrigan, 2003).

While the names and numbers of categories differ in each of the studies above, they provide an important insight into digital divide research. They highlight how, just like Internet use, Internet non-use is also not just one monolithic category, and does not straightforwardly remain on one side of the use/non-use dichotomy (Lenhart & Horrigan, 2003). It also provides an important critique of the common 'deficiency' discourse about non, or limited use of technology. As Selwyn suggests, drawing on Bauer (1995), non-users are sometimes viewed as 'abnormal' (Selwyn, 2003); despite the fact that not all non-users are disadvantaged and want to become users (Wyatt, 2005); and the Internet is not straightforwardly a 'good thing' for everyone (Eubanks, 2012; Selwyn, 2003; Wyatt, 2005).

This chapter is focused on exploring the experiences of young people who classify themselves as ex-users of the Internet. The original impetus for this research was the intriguing finding consistent in two nationally representative surveys in Britain that while internet use was ubiquitous for school aged children, around 10 percent of young people aged 17–23 described themselves as ex-Internet users. That is, it appeared that once some young people left school, they "stopped" using the Internet. The motivation of the study was relatively simple, namely to find out why this was the case and to investigate the implications that not using the Internet had in their lives.

METHODS

We conducted in-depth interviews with 22 young people in Britain aged 17 to 23. They were selected for the study because when asked the survey question: "Do you yourself personally use the Internet on whatever device at home, work, school, college or elsewhere or have you used the Internet anywhere in the past?" and were given three response options ("Yes. Current user", "No but I used it in the past", or "Never used the Internet"), the young people we interviewed selected the middle option, saying "No but I used it in the past". This question is identical to the one used to measure Internet use, ex-use or non-use of the Internet in Oxford Internet Surveys (OxIS)[1].

Data Collection

The interviews were semi-structured, carried out by one of the two authors, and lasted an average of 40 minutes. The focus of the interviews was quite broad, and took a semi-biographical approach (Bakardjieva & Smith, 2001), asking interviewees about their lives, quality of Internet access, motivations, skills and support to use the Internet, why they described themselves as an ex-user, if that mattered to them, and why. To recruit people for the study, we used a wide range of gatekeepers. These included parent groups, umbrella youth organizations, libraries, charities, colleges and schools.The sample consisted of 10 women and 12 men, who as is fairly typical of their age group, were in a phase of transition (Coles, 1995). There were those who were homeless, or in temporary housing; those who had recently become parents; young people who were unemployed,

---

[1] OxIS is a multi-stage national probability sample of 2000 people in Britain. Undertaken every two years since 2003, it surveys users, non-users, and ex-users, covering Internet and ICT access and use, attitudes to technology, and supporting demographic and geographic information. See http://oxis.oii.ox.ac.uk/.

or in part time or temporary jobs; and immigrants and refugees who had recently (typically in the past year) come to the UK, and who were trying to start a new life. As is clear from the description, the majority of this group was currently on low incomes—although this was not a criterion for participation.

Data Analysis

The interviews were audio-recorded and transcribed prior to analysis. The data were analysed thematically and iteratively to test and refine the categories (Richards, 2014). We found the framework proposed by Selwyn (2004), specifically theoretical access, effective access, use of ICT, meaningful use of ICT, and outcomes helpful in guiding our analysis. In line with Satchell and Dourish, we tried not to think of ex-users as somehow an issue that needs to be addressed on the road to technical adoption, but more that exploring ex-use needed to take a more holistic perspective, that took in the wider "cultural milleu". Such a holistic perspective allows a more complex analysis of how people are positioned (or position themselves) in a digitally mediated society (Satchell & Dourish, 2009, p. 9).

FINDINGS

When we asked participants the OxIS question about Internet use outlined above, the messiness of the boundary between use and non-use of the Internet was highlighted. As Kate (who ultimately decided she was an Internet user[2]) demonstrates:

Kate:	um…it depends like, when you say no…. how far along is no?
Int:	That is a very good question...so it is how you feel really?
Kate:	Yes—I use it. I am yes, but I am almost no...
Int:	and why is that?
Kate:	Because of circumstances (...) because of not having a laptop or Internet at home which is where I am the majority of the time, so having to go to the library or to the Internet café.

For Kate, a lack of good quality access to the Internet, a core dimension of digital inclusion, almost tipped her into the ex-user category. As will be detailed below, similar to researchers such as Selwyn (2004) and Haddon (2005), we have found that non-use and use of the Internet is not easy to define. Indeed, all the young people (apart from one) in our study who described themselves as an ex-Internet user did to some extent technically "use" the Internet.

For example, when we asked Graham if he used the Internet he told us, "No but I used it in the past", because, "I only ever used it at school really. At home I have never been able to access." He considered himself an ex-Internet user since leaving school a year earlier. However, since leaving school he had started using

---

[2] Kate's story is not included in the remainder of the analysis as she defined herself as an Internet user.

Facebook on occasions—i.e. once every few weeks—because "in school I saw all my mates but after school I saw them less and less (...) with Facebook I can still talk to them."

Similarly, Jack responded, "I am not an Internet user." When asked why this was, he explained "I use it very rarely, maybe once a month." For him, he had not been an Internet user since the age of 14 when he had begun to be seriously bullied after setting up a Facebook page. He used the Internet to check emails once a month for details of the football matches he refereed.

In our group of interviewees, these young people typically used the Internet in very narrow, infrequent and, as we will see below, often unsatisfactory or unsuccessful ways. Email, Facebook and job searches were the most popular activities. For example, Anna only went online to "check her housing application." For Jeff "It's just e-mails that I check, really (...) you got to have Internet to check e-mails, right?" Simon only used the Internet to open his Facebook profile that a friend had set up for him: "when I open Facebook, if I see my friend, I chat [to] him. If not, I close it and I am out."

Thus, while these young people did use the Internet in an objective sense, we do not (and they clearly do not) see this kind of use as amounting to particularly meaningful or satisfactory interactions with the Internet. There is a difference between the subjective experiences of using the Internet versus whether someone objectively uses it or not. For us it raises the issue of what being an Internet user means, and what implications this has for policy and practice. We will return to this issue in our discussion of meaningful online interaction below.

Of the 22 participants we spoke to, there were complex and multiple reasons why they chose to put themselves in the ex-Internet user category, and each young person possessing their own unique combination of reasons. From the analysis of data there were two overarching stories that emerged.

1. Young people who used to be Internet users, but no longer felt they were—either because they were using the Internet a lot less in terms of frequency and / or breadth of use than in the past or were using the Internet less but in quite different ways (often moving away from social and casual uses towards more economic and instrumental ones).
2. Those who had never really felt like Internet users in their entire lives.

For example, Leah fell into the first group. She told us, "I really have no use for the Internet at all." She had stopped using Facebook two years earlier; when she switched her college course to hairdressing from a more traditional academic subject. For her, this change meant she no longer needed to use the Internet for college; and she preferred "real life" experiences and "getting out and about" over using the Internet. A lack of private and personalized access to the Internet at the time we spoke to her made her even less likely to use it, but for her, she did not really feel she was missing out by not using it, particularly as her friends did not really use Facebook.

Leo, who is working in telesales, said "I don't really have a use for it, like on a day-to-day basis (…) as I've grown up I've become less needing of it." Leo saw the Internet as primarily a social thing, but due to changes in his life circumstances and as part of growing up, he no longer saw the Internet as important for social activities, and now has a much more functional and targeted approach to using the Internet for helping him to achieve his different goals—finding a permanent home, getting a good job etc. Another important shift for him was moving from having Internet access on a home laptop to accessing the Internet on a mobile.

In this group issues of time and their life trajectory are important. Indeed, Murdock (2002) has developed the concept of a "technological career", suggesting that people's relationships with and use of technology over time may change. It is therefore crucial to see these factors not as static, but as constantly changing as a result of what else is happening at the same time.

While for the young people in this category there was a time in the past when they did consider themselves to be an Internet user, there were others we spoke to who had never really felt like that. This was often due to issues around access and skills to use the Internet. For example, Karen, who works as a cleaner, told us that she just "didn't get" the Internet. She had never had Internet access at home, and her interactions with computers and the Internet at school were extremely limited. She had never used MSN or Facebook, although did and still does enjoy playing games. Karen was aware of how different she is in relation to her peers, but felt like it is too late to learn and was ambivalent about the extent to which it really mattered in her life.

In both groups, the young people we spoke to were well aware of how different they were to the majority of their peers. As Nick told us, "I am just one of those very, very few who does not use it."

DISCUSSION

In the discussion below, we develop how and why these young people define themselves as ex-Internet users, exploring issues of: 1) the factors that help to develop a feeling of connectivity, 2) defining what people believe they are connected to, and 3) the meaningfulness of using (or not using) the Internet.

A Feeling of Connectivity

From our data we found three inter-related aspects to understanding how someone feels connected to the Internet. A feeling of connectivity here is defined as a sense of control or agency over the possibility to use the Internet, and can be seen as the set of "pre-requisites" a person needs in order to feel able to connect. While the whole is greater than the sum of the parts, the key factors are: quality of access, perceived level of skills, and the availability of proxy use.

Access

Access is a well-recognized issue in relation to digital inclusion. Increasingly this is not about any kind of access, but is becoming far more about the quality of that access (e.g. home Internet access and personalized access) (Zillien & Hargittai, 2009) or smartphone only access (Ofcom, 2016).

For many of the young people we spoke to access was a problem, in the sense that they did not have home and / or personalized access to the Internet. Some did have access via their mobile phone—but not all used it or saw this as using the Internet—which we will discuss more below. A lack of quality of access had different implications for the ex-users we spoke to in terms of their feelings of connectivity.

The majority of those who defined themselves as an ex-Internet user because they were using it less (i.e. in the first group above) tended to have experienced a change in the quality of access to the Internet around the same time as a change in life goals and circumstances, such as leaving college or having a baby. The extent to which no longer using the Internet as much was a specific decision or simply something that had simply happened to them varied within this group, as did the feelings they had about the extent to which it mattered.

For example, Susan was quite ambivalent, "I barely get a chance to go on the Internet, so it's like if I did have a chance to go on the Internet I'd probably check them [emails and Facebook messages] all the time. But it doesn't bother me anymore. I'm not at college so there's nothing really like to go on." New mum Hannah meanwhile, "didn't have time for it [the Internet] really" and took the Internet off her phone package "because there was no need for it".

For others in this group the restricted availability to the Internet meant that they prioritized their use. For example, Luke told us "Facebook, it was good like first year of college, it was alright, but now that all my work is just on the Internet and like the only time I get to use the computer it has to be for work." He still used Facebook, but just on his phone, but he did not go on the Internet for anything else unless it was work related, and he did this at college. Such targeted use is similar to Robinson's study of US high school children, which showed similar targeted strategies on homework and information seeking activities for school children who had to use public spaces to access (Robinson, 2009).

Among those young people who fell into the second group and had never really felt like an Internet user, there were some young people who were potentially able to access the Internet on their mobile phones, while others were only able to access the Internet in public spaces (e.g. Internet cafes, computer room, library). Quality of access was not a significant issue for those in this group who did not have the skills to use the Internet in any case, but for those who did consider themselves to be able to use the Internet, this poor quality access was a core reason for not using it and was often keenly felt.

Overall, such stories relate closely to Selwyn's motion of perceived (or effective) access. As Selwyn notes, "any realistic notion of access to ICT must be defined from the individual's perspective" (Selwyn, 2004, p. 348). While in theory, all of these young people had places where they could access the Internet, they found the lack of convenience, lack of privacy lack of time and / or lack of control over these public sites of access to have a significant influence on the ways in which, and extent to which they used the Internet.

Skills

A closely related issue is skills. The majority of our participants who were using the Internet less than in the past were reasonably confident in their skills. As Anna who had just had a baby told us, "Yeah, I'm alright on a computer…I can do the basic stuff". These kinds of self-declared statements were also supported by the way they talked about using the Internet in the past. For example, Hannah told us "Yeah. We could always find them [proxies]… I was quite nifty on the Internet, to be honest; you could always find a way around [firewalls at school]." Thus, while skills are often an important reason for digital exclusion (see e.g. Hargittai, 2010) for this group, a lack of skills was not a significant factor in their ex-use. In contrast, a significant number of those who had never really felt like an Internet user typically had problems with their skills. For example, Tasmin who was currently looking for a job, but planning to go back to college in the autumn, said she found the Internet "Just all mumbly-jumbly like if I want to search for something I put into Google, but sometimes like the wrong thing will come up." Similarly, Kelly told us, "I'm not too sure [about if I can send an email] because, like, if I go in front of a computer and, like, if I recognize, like, the envelope or something like that, I'll know what it is (…). You know when you scroll over with the mouse and then the mouse comes up and it tells you, like, "Okay, this is to send a message," or "This is to attach something" or "This is to do this," or (…) then I think I could." We suggest that without skills it is not possible to really use the Internet in a meaningful way and was a significant reason for a number of participants in the second group to consider themselves as ex-Internet users (Eynon & Geneits, 2016).

Proxy Use

While many of the ex-Internet users we spoke to did have someone who could use the Internet on their behalf if they really needed to, for some ex-users, particularly those who had never really felt like an Internet user, proxy use, that is the use of the Internet through another person, was crucial. For example, Nathan told us he had never really used the Internet himself, and had always tended to ask others to use the Internet for him. This was because "The thing about it is with me, I'm a person that if I know how to use something I need to know everything, the whole lot. So if I don't know if I'm just like, no, I don't know it, you deal with it." This will be discussed further below.

Who Feels Able to Connect?

Based on an individuals' theoretical and actual access to use of the Internet, combined with their level of skills and availability of proxy use we can begin to see important differences between these groups of ex-users emerging.Of those participants who had never really felt like Internet users, most did not feel connected to the Internet, i.e. they did not really have a sense of choice or control over their (non) Internet use. This is because of limited skills which meant having access or not was almost irrelevant or because despite having some skills the participants had no meaningful access to the Internet. Interestingly, the only participants in this group who felt connected were those who, despite limited skills, had high levels of proxy access to the Internet. It was this proxy access that made the difference between feeling connected to the Internet (despite their lack of direct use) and those who did not. For example Tasmin told us that it did not matter so much that she did not use the Internet directly, because through her proxy use it was "still like telling me what I want to know." Similarly Ryan did feel he was aware of what the Internet was, but did not use it himself. He told us, "anything I want to know people can kind of really find out for me."

For those in the first group above who felt like they were no longer Internet users but had been in the past, the majority did feel connected to the Internet, in so far as they felt they could access the Internet if they wanted to. The only participants in this group who felt less able to do so were those who felt their skills were relatively poor. For the others, their theoretical and effective personal access to the Internet made a difference to whether they in fact used the Internet at all and how they used it, but regardless of this actual use, they felt they could if they wanted to. For a few, no theoretical or effective personal access to the Internet tended to be more of a certain kind of choice (i.e. they describe this as something that they could change if they wanted to—although we would not suggest they can be considered 'refusenicks'). However, for others, a lack of theoretical personal access was something that simply came about and was not an active decision. For others, they still had some form of theoretical personal access (typically on a mobile), so had personal access in theory but did not see a mobile in the same way as using the Internet on a laptop or desktop and for them this access varied in its meaning.

Beliefs About the Internet

As noted above, exploring (non) use of the Internet requires a longitudinal perspective. Goode (2010) suggests examining technology use in the context of adolescents' life stories and identities, within the conceptual framework of 'technology identity', using identity as a theoretical and methodological guide to examine and explain the digital divide. Viewing identity, among other factors, as a product of participation in communities, Goode suggested that experiences of using the Internet in the past influence adolescents' relationships with technology today. Thus, in order to understand why this group of young people felt they were no longer using the Internet, a key issue is to understand what the Internet was to them. These beliefs were shaped both by their prior experiences of using the Internet and also the kind of engagements and support people had while using the Internet at school and home, which we consider below.

Previous Experiences

For those who had never really felt like an Internet user, their experiences of the Internet tended to be fairly limited, particularly for those with low Internet skills, both in terms of the range of activities they had undertaken online (most typically social networking and information seeking) and the amount of time spent online when they were growing up. Some heavy proxy users had used the Internet through others for a wider range of activities (e.g. shopping and job seeking) and those with higher skills had used the Internet at least occasionally for this slightly wider range of activities, but in general prior Internet experiences for this group were relatively narrow.

This narrow experience had quite significant implications for what these participants felt the Internet could offer them, as many were not aware of the full range of possibilities. For example, Graham saw the Internet primarily as a place for socializing and connecting with friends. Tasmin was not entirely sure what the Internet was and described the Internet "as a chameleon". Karen, felt the Internet was a "person that was really cocky" and while her friends had told her that you "can use the Internet for everything. There's, like, news, sport, lotteries online, being able to order things online." She did not know, when we were discussing the use of the Internet for job seeking, that it was possible to send an email attachment (in this case a CV).

For those who had felt like an Internet user in the past, they tended to have experienced using the Internet for a slightly wider range of purposes, alongside social networking, more entertainment activities and for school or college work were the most commonly cited and participants often mentioned at least one other activity such as information seeking, shopping, job seeking or house bidding. However, the breadth of activities for each individual was still relatively narrow relative to many in their age group (e.g. Livingstone and Helsper, 2007). Interestingly, the two participants who were perhaps closest to being an Internet user (who had significantly changed the way that they used the Internet) but still (to us at least) seemed to use it and had the widest previous experience of engagement with the Internet of all our participants and continued to do so.

It is worth noting though that what the Internet was also changed for some participants depending on the device being used. For those who accessed the Internet on their mobile they rarely saw this as "using the Internet", instead they were using their phone for specific tasks. As Luke told us, what talking about his use of Facebook:

Luke:	Because it's literally, it's just an app. So just Facebook. So I just click onto it and I'm on Facebook.
Interviewer:	So that doesn't feel like the Internet to you?
Luke:	Not really you know.
Interviewer:	It's a different thing?
Luke:	Yeah. Because Internet to me is like just going onto Google, and then you know when you type in what you want and press enter and then all what comes up, that's the Internet to me and that's where I find it tricky.

Support Networks

These experiences of using the Internet were very much shaped by the support provided by others (e.g. parents, friends and school) that were on offer to our participants when they were younger. In general, few people who defined themselves as never using the Internet had had any kind of support in learning to use the Internet either at school or at home. If school did offer support, it tended to be more based around IT and word applications, as opposed to the Internet. None of the participants we spoke to in thus group had ever had Internet at home (although a couple did have access to a computer) or parents or siblings who could help them. This lack of support contributed to a limited awareness of what the Internet could be used for. The main source of support for people in this group tended to be their friends, although for some this led to further problems. For example, Reena's main experience of using the Internet as a teenager was for MSN with her friends. No one in her family used it. This meant that no one could share additional expertise, as she explained, "I think it's [the Internet] quite useless for some people who don't know much about it. ... Like, my mum and dad, they wouldn't know what to do if they were on a computer." Without a sufficient range of experiences and skills, family and peer networks do not take people very far in their Internet use and their development of skills (Eynon & Geniets, 2016). For those in the first group who felt like they used to use the Internet, they tended to have experienced a little more support in the past, although this tended to operate in slightly different ways for each individual. Typically, the more support and better access a participant had experienced in the past (either at school and or at home) the more confident they were that they could use the Internet again in the future if they wished to do so.

Thus, similar to Lee we would suggest that, "inclinations and opportunities [to use the Internet] are shaped at the level of the school (as a community and space), the household and the individual" (Lee, 2005, p. 320). Considering the interactions between these three levels and the Internet are crucial.

Meaningful (Non) Use of the Internet

The majority of the young people who used to feel like Internet users but no longer did, told us how the Internet was not particularly meaningful to them. This was typically because the Internet did not really offer them much in terms of achieving current needs or goals. As noted above this group of young people primarily saw the Internet as able to support communication, school / college work and occasionally, for some participants, entertainment, information seeking, job seeking and / or purchasing goods. BBM or text was often used to communicate with their friends and not all were fans of Facebook. Leo did want to be a person who just spent their time "sitting on Facebook." Hannah, had got tired of what Facebook offered her, she told us, "I don't want to see that some person I went to primary school with is eating toast. I don't want to see things like that," (…) I've kind of got better things to do."Most, like Susan, were no longer at school or college and this meant less use of the Internet. The few of our participants still at college, like Luke, tended to prefer books, handwriting his assignments and described a need to be "pushed" to use the Internet, saying he "would go around it" where possible. For him, "I wouldn't really sit down if I never had no work and then use the Internet, because I wouldn't really know what to use it for." In terms of entertainment, many preferred listening to CDs or watching TV. For example Jai saw the Internet as something to pass time, but found watching TV fulfilled this purpose better. Jasmine, who watched TV and listened to the radio, did not feel like she was "missing out" by not using the Internet. And Nabeel felt the Internet was "running out of ideas" and "not offering me as much as it should."

Sources of information other than the Internet were often considered better and more reliable: Jai liked to go to the library and read books, Anna preferred her mother and baby group for information on being a mum. Susan talked about her loves of writing poetry and reading, but preferred library visits and local classes to online forums. A number of the young people pointed out the problems with finding information they could trust. While some did see the possibilities of using the Internet for job seeking, few felt it had a lot of worth. Most of the young people preferred dealing with other people face to face. Jai pointed out that while the Internet could make it easier to search for jobs, it did nothing to change the fact that there were hardly any jobs for young people like him available. "I'm meant to be searching for jobs that are related to my CV, like what type of qualification and experience I've got on my CV. Half the time I don't even do that, because half the time when I do do that, there's not really any jobs for me. So then again, that pushes you away to keep searching [online]."

Financially, online shopping was not often a realistic option as few had the resources to do so. However, the occasional instance was mentioned. For example, Jasmine told us she last used the Internet about 6 months earlier when she signed up for her pregnancy voucher pack online, "the bounty pack" and printed off a "voucher to get a discount on pushchairs."

Many in this group had relatively ambivalent and sometimes quite negative views of the Internet. For example, Anna, just did not "think the Internet's that brilliant a thing, to be fair." Hannah was quite disparaging of the gossipy and nosey nature of the Internet, pointing out that "As soon as you log onto Facebook the first thing it says was, "What's on your mind?" and "all Twitter is is just seeing what celebrities are breaking up and I'm not into that. Any main news you can see on TV." Yet, they did have some positive experiences to. For Hannah, who used her mobile phone before the birth of her baby to look up information about childbirth, "I used to read up on, like, birth and things like that. I think most probably in my whole life that's probably the best thing the Internet has helped me for."

Thus, it is reasonable to suggest that the majority of this group did not have a strong sense of connectivity, they were not that impressed with what the Internet could offer them in the current context of their lives, and overall their relatively infrequent and targeted use of the Internet was not particularly meaningful. Yet, interestingly for some their (non) Internet use was, to some extent meaningful in the sense that it helped to define who they were. For example, Hannah was "more of a get up and go person." For Jack, who had been heavily bullied, his control over his (non) Internet use was particularly important. This has been found in other studies of non-use. As Sewlyn suggests, drawing on de Certeau (1984), sometimes non- use is a way to achieve some kinds of power for groups that are rarely able to express forms of power (Selwyn, 2003).

This is not to say that everyone in this group felt the same way. For another minority, who experienced problems with private access or were still not confident in their skills, or who felt their access to the Internet was not sufficient to achieve their goals, their lack of use was also meaningful. Josh told us "I do find it hard. If I had Internet access every day I reckon I could have a job within probably a week or two."

In the second group, who had never felt like Internet users, more of our participants felt that their (non) Internet use was more often meaningful in the sense they felt they were missing out to some extent by not using it, typically for job search, information and connecting with friends because they did not have the necessary access or skills. It was also at times meaningful in a sense that our participants felt different from others. Kelly considered herself as different from her friends. She told us "I've never been, like, a computer person" and later "I'm the one that doesn't really know about computers more than anyone else." Karen told us how she felt "really old-fashioned." For young people like Karen and Kelly, their lack of use was typically felt in social terms. For example, Karen felt her social circle was shrinking: "I don't have a lot of friends now, it seems, because I've been forgotten about because I don't have all these little network things that they do." While she "wouldn't say it's like a massive deal" it was "still kind of upsetting, I suppose." While these participants did, in some senses at least want to learn to use the Internet, they felt it was too late. As Kelly explained "I would love to learn....(...)...I've even, like, tried to go on an ICT course but now they've said that I have to pay for it because I'm 19. For Karen, "I feel like I've gone past the stage of trying to learn it now and I would rather just have to do it... you know, live without it."

For those participants in this second group who were proxy users, (non) use of the Internet tended not to matter a great deal, if at all. Ryan told us, "me personally, I'm fine without it". Nathan told us that when he was younger "he didn't really see the Internet as a good way of spending his time" and this seems to still be the case at 22. It makes sense for him for others to use it for him as they are "more fluent". For Tasmin, her proxy use was sufficient in most cases. However, she is wary of her poor skills and talks about how she tried to set up a Facebook account—but "just couldn't finish it" because it was complicated. She told us "I wasn't able, I didn't get it and I don't like to admit because I think, "Oh am I the only dumb one, because everyone else has got one". But I just find it complicated and I thought, "Forget it"."

Selwyn (2004) highlights the important distinction between use of ICT versus meaningful use of ICT, "where the 'user' exerts a degree of control and choice over the technology and its content, thus leading to a meaning, significance and utility for the individual concerned (Bonfadelli, 2002; Silverstone, 1996)" (Selwyn, 2004, p. 309). In our groups, we observed that the meaning, significance or utility in their (non) use of the Internet varied over time context and individual. In some cases (non) use of the Internet, was not in their terms, a problem. However, we would suggest that this does not straightforwardly mean it is not a problem in the UK where many services and support for young people are moving online. At other times their (non) use of the Internet is keenly felt, either in terms of identity or in achieving life goals.

DISCUSSION AND CONCLUSION

In this qualitative study we have highlighted the existence and complexity of ex-use of the Internet among young people—a group where high levels of digital engagement tend to be taken for granted. Globally, the UK ranks reasonably highly on digital inclusion indicators, with almost 9 in 10 adults in the UK saying they use the Internet in some capacity (Ofcom, 2016). Yet, this study shows that even within countries which are relatively digitally enabled, there is still uneven engagement, and despite the popular assumption that exists across countries, not all young people use the Internet in meaningful ways.

In our work we distinguished two broad groups of ex-users. One of whom perceived themselves as no longer using the Internet, primarily due to a significant change in access, and / or a change in the amount or nature of use. These changes in use that our participants describe are notably very similar to the ways that Internet use is typically measured in terms of access to the Internet, amount of time spent online, the breadth of Internet use, or the nature of use (Selwyn, 2006; Van Dijk, 2006). A significant change in one or more of these aspects tended to make the young people we spoke to feel like they were no longer "proper" Internet users. Our second group had never really felt like they used the Internet in the first place, often due to a complex interplay of factors between access, experience, proxy-use, skills, and support—factors also well explored in digital exclusion research. Indeed, our research has resonance with an array of digital inequality literature from other advanced market economies.

Interestingly, alongside time as important element, many of our participants defined their ex-use by delineating themselves in contrast to their peer group, and this social element is important. What it means to use or not use the Internet may be different for younger generations who have grown up with the Internet as part of their everyday lives compared to the meaning ex-Internet use has for older generations.

The complex array of factors outlined above that help us to begin to understand ex-use of the Internet and the meaning this has, demonstrates that there is a strong inter-relationship between the individual and the social context which cannot be separated out or underplayed (Halford & Savage, 2010; Warschauer, 2004). The classic Sociological question of agency and structure reverberate throughout this data. As Lee (2005) notes in her study of young people's uses of the Internet, "Inclinations and opportunities are multidimensional, shaped by a range of factors relating to individuals' place in wider social structures and their personal situations, as well as their perceptions and experiences of the Internet" (p. 318). As authors suggest, such experiences need to be explored and understood over the life-course and not treated in a cross-sectional way (Murdock, 2002).

The young people in this study are not the same as other ex or non-user groups who are performing resistance from a relatively strong social position, where non-use of a particular platform is more of a choice (e.g., Portwood-Stacer, 2013) or where non-use is more based on a choice informed by the utility of the platform (e.g., Birnholtz, 2010). They are different both because the participants discussed here are largely non or ex users of the Internet across all applications—not just one (Wyatt, 2005); and because many of them are currently experiencing complex social arrangements with a lack of material resources. Their ex-use perhaps reflects an understanding by these young people that the Internet does not hold any particularly miraculous path to prosperity that is sometimes promised to them (Eubanks, 2012). Indeed, they are highly reflexive of their current social conditions (Davies & Eynon, 2018). Thus, moves to support these young people need to be made within a wider commitment to addressing social as well as digital inequalities (Warschauer, 2004). This issue is core for all global leaders interested in addressing issues of digital inequality. Too often the digital becomes the panacea for all forms of social injustice, yet the research here and the work cited highlights time and time again that this is not the case. In a digitally enabled world, technology is important, yet it is only one factor among many.

A second important issue for global leaders is to use qualitative work such as this to inform the development of future survey instruments used to measure issues of digital inequality. We would echo other researchers that have argued for digital inclusion surveys to allow for a scale of very limited frequency and breadth of Internet use, when conceptualizing use and non–use of the Internet (Haddon, 2004; Livingstone & Helsper, 2007; Murdock, 2002; Wyatt, 2003); to move away from a binary distinction between use and non-use (Barzilai-Nahon, 2006; Halford & Savage, 2010; Robinson, 2009).

Based on our findings we would also suggest that it might be beneficial to move away from the primary focus on measuring Internet use towards more attention to outcomes of this use (Van Deursen & Van Dijk, 2014; Van Deursen, & Helsper, 2015; Eynon, Rauer, & Malmberg, 2018). Similarly Allen, developing some of the ideas proposed by Jung et al. 2001, suggests researchers focus on the experience of connectivity as opposed to Internet use, to better understand the social implications of the Internet. He argues that "the experience of connectivity means the way people utilize the Internet to achieve a variety of outcomes in their everyday lives. This experience includes both the importance people place on connectivity for successful achievement of outcomes and also the range of outcomes which it influences" (Allen, 2010, p. 351). A more comprehensive move away from questions on Internet use to those focusing more on meaning and connectivity, alongside greater recognition of the wider social structure could be highly valuable for evidenced based policy making.

Measurement has of course always been a complex political act, as they shape the way that we think about and understand the world (Espeland & Stevens, 2008). Measuring digital inequality is also an ever changing practice that needs to be understood within a particular timeframe and national context (Reisdorf, et al., 2017); and care must be taken when making global comparisons. Perhaps given the pervasive but inaccurate discourses around young people and technology, and increasing concerns (in the UK at least) about increasing inequalities more generally, there is even more need to ensure that digital inclusion researchers and policy makers think about and employ a range of measures that capture as best as we are able a more refined understanding of ex-use of the Internet that can be of value both theoretically and practically. This seems of particular relevance in light of the ongoing changes of the knowledge economy (Normore & Ilon, 2010), to ensure that this group of young people are not further excluded and left behind. Regardless of country, this need for a more nuanced perspective is essential for global leaders to develop meaningful policies and interventions in this area.

ACKNOWLEDGMENTS

This study was enabled through a grant from the NominetTrust. The authors would like to thank all the young people and the youth organizations that contributed to the research by allowing the authors to interview the young people on their premises.

REFERENCES

Allen, M. (2010). The experience of connectivity. *Information, Communication and Society, 13*(3), 350–374.

Bakardijeva, M., & Smith, R. (2001). The Internet in everyday life: Computer networking from the standpoint of the domestic user, *New Media and Society, 3,* 67–83.

Barzilai-Nahon, K. (2006). Gaps and bits: conceptualizing measurements for digital divides. *The Information Society, 22,* 269–78.

Bauer, M (1995) (Ed.), *Resistance to new technology.* Cambridge UK: Cambridge University Press.

Baumer, E. P., Ames, M. G., Burrell, J., Brubaker, J. R., & Dourish, P. (2015). Why study technology non-use? *First Monday, 20*(11). Available at, https://ojphi.org/ojs/index.php/fm/article/view/6310 [accessed 1st December 2016]

Birnholtz, J. (2010). Adopt, adapt, abandon: Understanding why some young adults start, and then stop, using instant messaging. *Computers in Human Behavior, 26*(6) 1427–1433.

Coles, B. (1995). *Youth and social policy.* London, UK: UCL.

Davies, C., & Eynon, R. (2013). *Teenagers and technology.* London, UK: Routledge.

Davies, C., & Eynon, R. (2018). Is digital upskilling the next generation our 'pipeline to prosperity'? *New Media & Society.* Retrieved from http://journals.sagepub.com/doi/abs/10.1177/1461444818783102

de Certeau, M. (1984). *The practice of everyday life* (S. Rendall, Trans.). Berkeley, CA: University of California Press.

Di Maggio, P., Hargittai, E. Neumann, W. R., & and Robinson, J. P. (2001). Social implications of the Internet, *Annual Review of Sociology, 27,* 307–336.

Espeland, W. N., & Stevens, M. L. (2008). A sociology of quantification. *European Journal of Sociology/Archives Européennes de Sociologie, 49*(3), .401–436.

Eubanks, V. (2012). *Digital dead end: Fighting for social justice in the information age.* Cambridge, MA: MIT Press.

Eynon, R., & Geniets, A. (2016). The digital skills paradox: How do digitally excluded youth develop skills to use the internet? *Learning, Media and Technology 41*(3) 463–479.

Eynon, R., Rauer, U., & Malmberg E-L. (2018). Moving up in the Network Society: A longitudinal analysis of the relationship between Internet use and social class mobility in Britain. *The Information Society, 34*(5).

Facer, K., & Furlong, R. (2001). Beyond the myth of the 'cyberkid': Young people at the margins of the information revolution. *Journal of Youth Studies 4*(4), 451–469.

Goode, J. (2010). The digital identity divide: How technology knowledge impacts college students, *New Media & Society, 12*(3), 497–513.

Haddon, L. (2004). *Information and communication technologies in everyday life: A concise introduction and research guide.* Oxford: Berg.

Haddon, L. (2005). *Personal information culture: The contribution of research on ICTs in everyday life.* UNESCO between Two Phases of the World Summit on the Information Society, St. Petersburg, Russia, May 17th–20th, 2005.

Halford, S.,& Savage, M. (2010). Reconceptualizing digital social inequality, *Information, Communication & Society, 13*(7), 937–955.

Hargittai, E. (2010). Digital na(t)ives? variation in internet skills and uses among members of the net generation, *Sociological Inquiry, 80*, 92–113.

Helsper, E. J., & Eynon, R. (2010). Digital natives: where is the evidence? *British Educational Research Journal 36*,(3), 503–520.

Jung, J-Y, Qiu, J. L., & Yong-Chan K., (2001) Internet connectedness and inequality beyond the divide. *Communication Research, 28*(4), 507–535.

Lee, L. (2005). Young people and the Internet. From theory to practice. *Nordic Journal of Youth Research, 13*(4), 315–326.

Lenhart, A., & Horrigan, J.B (2003). Re-visualizing the digital divide as a digital spectrum. *IT & Society, 1*(5), 23–39.

Livingstone, S., & Helsper, E. (2007). Gradations in digital inclusion: Children, young people and the digital divide, *New Media and Society, 9*(4), 671–696.

Murdock, G. (2002) *Tackling the digital divide: evidence and intervention.* Paper given to The Digital Divide Day, Seminar; 2002 Feb 19. Coventry: British Educational Communications and Technology Agency.

Normore, A., & Ilon, L. (2010). Globalizing educational leadership for innovation and change, *International Journal of the Humanities*, 8 (2), 12–140.

Ofcom (2016). Smartphone by default internet users. London, UK: Ofcom. Retrieved from https://www.ofcom.org.uk/research-and-data/telecoms-research/mobile-smart-phones/smartphone-by-default-2016 [accessed 7/12/17]

Portwood-Stacer, L., (2013). Media refusal and conspicuous non-consumption: The performative and political dimensions of Facebook abstention. *New Media & Society, 15*(7), 1041–1057.

Reisdorf, B. C., & Groselj, D. (2017) Internet (non-) use types and motivational access: Implications for digital inequalities research. *New media & society, 19*(8) 1157–1176.

Reisdorf, B. C., Triwibowo, W., Nelson, M., & Dutton, W. (2017). an interrupted history of digital divides. *AoIR Selected Papers of Internet Research*, 6. Retrieved from https://spir.aoir.org/index.php/spir/article/view/1256 [accessed 1st December 2016]

Richards, L. (2014). *Handling qualitative data: A practical guide.* Third Edition. London Sage.

Robinson, L. (2009). A taste for the necessary, *Information, Communication & Society, 12*(4), 488–507.

Satchell, C., & Dourish, P. (2009). Beyond the user: Use and non-use in HCI. In *Proceedings of the 21st annual conference of the Australian computer-human interaction special interest group: Design: Open 24/7* (pp. 9–16). ACM.

Selwyn, N. (2003). Apart from technology: understanding people's non-use of information and communication technologies in everyday life. *Technology in society 25*(1), 99–116.

Selwyn, N. (2004). Reconsidering political and popular understandings of the digital divide. *New Media & Society, 6*(3), 341–362.

Selwyn, N. (2006). Digital division or digital decision? A study of non-users and low-users of computers, *Poetics, 34*(4), 273–292.

van Deursen, A., & Helsper, E. (2015). The third-level digital divide: who benefits most from being online? Communication and information technologies annual: digital distinctions and inequalities. *Studies in Media and Communications 10*, 2–53

van Deursen, A., & van Dijk, J. (2014). The digital divide shifts to differences in usage. *New Media & Society, 16*(3), 507–526.

van Dijk, J. (2006). *The network society, social aspects of new media* (1st ed.). London, UK: Sage.

Verdegem, P., & Verhoest, P. (2009). Profiling the nonuser: Rethinking policy initiatives stimulating ICT acceptance. *Telecomm. Policy, 33*(10–11), 642–652.

Warschauer, M. (2004). *Technology and social inclusion: Rethinking the digital divide.* Cambridge, MA: MIT Press.

Wellman, B., & Haythornthwaite, C. (2002). The internet in everyday life—An introduction. In B. Wellman & C. Haythornthwaite (Eds.) *The internet in everyday life* (pp. 3–45). Oxford, MA: Blackwell.

Wyatt, S. (2005). Non-users also matter: The construction of users and non-users of the Internet. In N. Oudshoorn & T. Pinch (Eds.), *How users matter* (pp. 67–79). Cambridge, MA: MIT Press.

Wyatt, S. M. E., Oudshoorn, N., & Pinch, T. (2003). Non-users also matter: The construction of users and non-users of the Internet. *Now Users Matter: The Co-construction of Users and Technology,* 67–79.

Wyatt, S., Thomas, G., & Terranova, T. (2002). They came, they surfed, they went back to the beach: Conceptualising use and non-use of the Internet, In S. Woolgar (Ed.) *Virtual society? Technology, cyberbole, reality* (pp. 23–40). Oxford, UK: Oxford University Press,

Zillien, N., & Hargittai, E. (2009). Digital distinction: Status-specific types of internet usage. *Social Science Quarterly, 90*(2), 274–291.

PART II

DIGITAL EQUITY AND ACCESS ISSUES

CHAPTER 4

LEADING THE COHORT ACROSS THE DIVIDE

Recent Best Practices to Enhance Cohort Teaching and Learning

Steven C. Williams

This chapter brings focus to the best practices of an inner-city university program which uses technology to enhance cohort teaching and learning. The chapter draws on the debates about the digital divide by showing how to leverage seven best practices that can be used to create an equitable learning space where cohort participants learn to overcome many core technology disadvantages to become actualized online learners. Implications for leadership will be discussed.

At California State University, Dominguez Hills (CSUDH), the College of Education has created the School Leadership Program (SLP) that is designed to develop and fortify skills among cohorts of K–12 school administrators in a continuous improvement teaching and learning environment. Serving over 17 school districts in the greater Los Angeles region including the nation's second largest school district, the Los Angeles Unified School District, the SLP has created a leading-edge online program utilizing a lean staff of less than 5 where more than 3,000 students have attended over the past decade. Since its move to an online

Crossing the Bridge of the Digital Divide: A Walk with Global Leaders, pages 59–78.

program in 2006, the SLP has worked closely with Dominguez's campus technology leaders to leverage innovative approaches that have addressed and mitigated some of the most pressing technology divides which confront the contemporary school administrator.

At the core of this approach has been the design and construction of course shells within the campus learning management system (LMS). The course shells not only provide content for the program, but also drive cohorts of students sequentially through the program in a timely and efficient manner. The program has made a conscious effort to structure staff and resources to maximize the course shell within the LMS to smooth entry into the program. It also provides students with technology skill development to cope with the program itself, to build and extend a sense of community within cohorts and, finally, to impart educational technology competencies on participants who graduate and leave the program. To that end, many participants return to their school sites mobilized to share some of that skill development and knowledge with their own staff and students.

The SLP's efforts really are an attempt to cope with what it conceives as the "digital divide" for members within its cohorts. To appreciate how the SLP understands the concept and how it applies to the program, it is important to unwind the term as it evolved through time. To that extent, there has been no shortage of contentious debate concerning concept of the digital divide. To understand its current meaning and how it affects students in the SLP, it is important to underscore the contours of the debate as it developed in the late 1990s. At the outset, much scholarly discussion involved at its core the inclusion of a multiplicity of causes related to class and race as some of the main determining factors in how the issue was defined and how it was to be addressed. Researchers have noted that "the ratio of students to instructional computers with Internet access was higher in schools with the highest poverty concentration (percent of students eligible for free or reduced-price lunch) than in schools with the lowest poverty concentration" (Parsad & Jones, 2005, p. 8). Essentially, the debates were about how to best spread "the future" as it were in the form of the educationally enlightened internet. The idea was to do it in a manner that appeared to the public to be as equitable as possible.

In those early days of the debate, important conflicts were often over policies that could mitigate actual hardware disparities between richer school districts and poorer ones. At the time, there was far less concern about how that hardware would be used at the software level than is the current practice. Instead, the push was to connect schools to the internet as fast as possible and, at that time, the main problems were mostly hardware related. Wiring the school site literally involved wire, routers, firewalls, desktop computers and an assortment of other technology infrastructure devices. The E-rate program of the late nineties and early "aughts" (a term used to describe the years between 2000–2010), informed by these debates over equitable hardware distribution, was one of the key components which emerged as the cornerstone of the education establishment (Federal Communications Commission, 2004).

To the extent that poorer school districts and the schools within those districts received more resources to address the hardware digital divides of the era, the factors which created those divides in the first place never went away. If anything, those issues became more acute depending on particular circumstances—and accelerated after the 2008 recession when especially acrimonious fights between the haves and have-nots in education broke out (Cohron, 2015; Kennedy, 2016; Mountain, 2017; Osborn, 2016; Paul, 2016; Reynolds & Chiu, 2016; Rogers, 2016).

In the wake of this tumult, newer digital divides emerged at the school site with added dimensions of insidious interactions occurring among them which—in its broadest contours—encapsulates our current predicament. As newer gaps emerged, the school site became a replay of the older divides introduced to newer divides. The following are examples of digital divides that have persisted, evolved and, in many cases, mutated into something far more concerning which tend to impact poorer schools more disproportionately:

- Hardware procurement and deployment where poorer schools are more often disadvantaged by uninformed purchasing decisions and sometimes shoddy installation[1]
- Properly vetted software selection
- Staff training
- In-classroom usage with students
- Here are newer divisions[2] which have emerged:
- Gender distribution
- Access to computer power / internet at home
- Accessibility
- Social media distraction culture

No longer were issues of equity fixated solely on numbers of computers in the school lab, but on a host of other dimensions as well, many of which were being driven by more powerful social and economic forces. Researchers have noted that when cross-matrixed, these factors can put schools at a true disadvantage when trying to leverage technology to construct a stable learning environment. A key study on the subject concludes in an understated tone that "the digital divide is a multilayered phenomenon" (Ritzhaupt, et al., 2013, p. 292).

When we look back, even as more and more school sites became equipped with the hardware basics, newer fissures emerged leading crucially to the endpoint of the entire endeavor: software usage in the classroom with students. Increasingly, observable divides have been noted by researchers in the ways in which various software are utilized as a means of enhancing teaching and learning. Sadly, the early studies on this phenomenon indicate a stark repetition of patterns detected during the hardware phase of the digital divide where schools from middle to

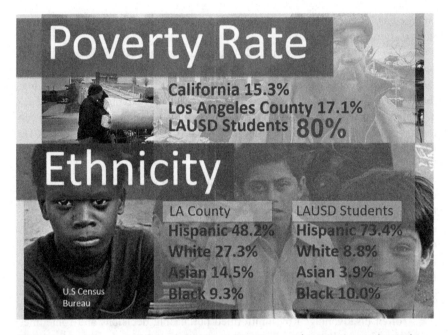

FIGURE 4.1. From an SLP PowerPoint presentation emphasizing several social issues which program directors and staff know all too well. "Case in Point: Challenges, Lessons and Progress," Ann Chlebicki and Antonia Issa Lahera. These statistics reflect the socio-economic conditions of schools from which the majority of SLP students work.

upper socio-economic income brackets tend to be more powerful software users not only as consumers of software products, but as producers too (Ritzhaupt, et al., 2013, p. 292). Moreover, several studies show that graduates of those K–12 schools where students can leverage hardware/software resources efficiently and effectively are more likely to go to better colleges, and attain better degrees and jobs (Herold, 2017). Hence, the digital divide—even in its recent mutations—shows repeated patterns of social inequity as was observed in earlier phases of its history. It seems, the past of the digital divide is never dead. It is not even past.[3]

The SLP which emerged at Dominguez Hills in this milieu has consciously sought to embed within its practices and its strategies of design a sense that nearly *all* participants in the program have some significant digital divide which must be hurdled for students to be successful in the program. Program developers view the entire cohort as a unit through which skills will be imparted and best practices will be actualized. Though there are many exceptions, the assumption that the SLP staff have in their design concept is that most students in the program work in school districts and schools usually considered, by current metrics, on the wrong side of the digital divide—the so-called Title I schools (Malburg, 2015) which have been impacted negatively by poverty, multilingualism, overcrowded conditions, and low teacher morale. See Figure 4.1.

Hence, skills and tech-knowledge acquisition in the classroom for many incoming SLP students have been circumscribed by the underdeveloped technological work environment of their own school sites. So, while most SLP students have a working knowledge of the basic of technologies of their own school sites, it is often not adequate for the rigors of the SLP. It is not that students who come into the program lack technical skills *per se*, but rather many of the skills they possess are not readily transferable to the requirements and necessities of the program. For many incoming students, programs like the SLP are compelled to reduce the shock of the new which can occur for students who come from technology-deprived school sites. Paradoxically, it is not obvious that new students may not have many technology skills beyond smartphone operation. One educational researcher summed up the false sense that students are skillful technology users just because they appear to know what they are doing on their smartphones:

> The digital divide is not always glaringly obvious. It is commonplace to see a classroom replete with laptops and students who have developed, in my opinion, an unhealthy attachment to their smartphones. [We] must be careful not to equate their cursory observations of students using digital devices with students having high levels of digital literacy and access (Bodrick, 2016, n.d.).

Other recent research has pointed out the growing generalized attention deficit disorder of students who cannot untie themselves from their social media networks like Facebook and Instagram. One study shows, on average people reach and touch their cellphones over 2,600 times a day (Winnick, 2016). The growing dominance of the "attention economy" at the school site and at home has become a real concern for educators competing for that same attention space. Many students are often driven to high levels of distraction caused by the near narcotic effect of the smartphone "notification" revolution. Recent studies have shown that a constant stream of notifications of likes and retweets can have a stimulating effect on endorphin production creating an almost instantaneous brain rush triggered by merely viewing notifications on a smartphone (Gazzaley & Rosen, 2016).

While the sheer time drain of social networking can be a major distraction from the intense content goals of the program itself, most SLP students enter the program as already mature adults with jobs, families and social obligations that extend well beyond the program. It would be an understatement to say that, generally, students in the program are busy people. Within this churn of life's day-to-day intensity both at the school site and at home, it is incumbent on the SLP staff to design and incorporate strategies to maximize time efficiencies, reduce frictions between content and the students, increase engagement so as to enhance motivation for the program itself—and do most of this in an online space.

To understand the process by which the SLP team approaches the development of content and strategies to disseminate that content in an effective manner to its students, it is important to understand the program's seven basic best practices by which the program operates and which fuels the program's success. From these practices emerge a recipe for how to leverage a lean staff into a large cohort program support structure. The practices which the SLP staff have developed represent a leading-edge way of doing online learning. The ease with which students interact with the program and its content undergirds its students' success.

Best Practice 1: Scale Your Issues to a Size You Can Manage

Avoiding redundancy of task became one of the first orders of business for the SLP. The program was born around the time at which CSUDH had introduced its learning management system (LMS) and redundancy of task became an immediate issue. Here is how: From the outset, the SLP offered its cohorts a number of courses during a given semester. As cohorts grew in number from year-to-year, sections were added to each course offering to accommodate newer entering cohorts. Over time, a given course might grow to 5 or more sections to meet the demands of multiple cohorts.

Suddenly, within the first few years of the program, multiple sections of multiple courses produced a crisis of content development and management. Here is an example of how a small program like the SLP can get subsumed by the overwhelming numbers of courses and course sections: In a recent semester where the multiplier effect for multiple sections of the same SLP553 Organizational Leadership course occurs, staff was responsible for seven different sections of the same course. Now factor in these additional courses, each multiplied by 4 to 7 sections depending on the number of cohorts and within the first few years of the program, the SLP's lean staff of 4 to 5 people suddenly became responsible for shepherding more than 50 courses shells in a given semester.[4] As the proliferation of multiple sections became apparent, the SLP faced a numbers crunch that almost immediately threatened to engulf the program.

Provoked by necessity, the SLP reached out to campus technology leaders to produce what now seems, in retrospect, a seemingly simple but elegant solution. The idea was to produce a pre-built course shell template for each course. Templates would be filled with highly vetted content and organized in sequential order which would then be copied in mass—using the semi-automated course copying mechanisms of the LMS—to the various sections of each course. So, instead of facing the daunting task of filling content separately for upwards of 50 course sections, instead the program concentrated on core template development for 7 or so basic course templates.

Under the normative processes of LMS content development of the era, this issue would not have been so pressing given that traditionally instructors are responsible for producing their own content. But the SLP took a decidedly programmatic approach to ensure that content development was consistent across all courses and sections to those courses. In fact, the SLP worked intensely in a team-oriented environment with a group of the program's instructors to create uniformity of content across all course master shells built for the program so that no matter who was assigned to teach a course, content development for that instructor would not be an issue. Once core content components were agreed upon and structured in the master course shells, instructors who taught in the program were now free to plug into pre-build LMS course sections. Alleviated of the burden of individuated content development, instructors would, instead, guide students through these pre-built materials focusing their responsibilities on student motivation, interaction and assessment—all necessary high-touch elements especially for students generally unfamiliar with an online learning experience.

The advantages to this method became apparent almost immediately. Singular core templates could be carefully crafted per each core course requirement of the program. And in turn, those templates could be rapidly copied and spawned to specific course sections representing different tracked cohorts. But beyond addressing the multiplier effect and reducing the stress of the numbers crunch, by creating a template-driven approach to content development, the SLP standardized its content development which allowed it to more fully align the goals and objectives of the program. Hence, the program embedded a consistency throughout all its courses and cohorts. Quality was more highly assured from course to course and throughout the cohorts. In large measure, this was due to the innovative use of the LMS's course copying tools.

Furthermore, staff members, now freed of most of the grind of repetitive and redundant task mastering, could more fully utilize their limited time resources on improving the user experience of master course shells. Though the template model proved an important innovation to that end, it still had several shortcomings that soon drove the team to search for other innovative solutions.

Best Practice 2: Don't Be Afraid to Invest In and Utilize Outside Technology Vendors If Your Own Technology Isn't Sufficient

The learning management system which the CSUDH campus supported at the time (and currently in 2018) is Blackboard. Blackboard has been a leader in the development of the higher education learning management system, and beginning in the late nineties it established its beachhead in several campuses across the nation including CSUDH. In some of the most profound ways, it was a prescient choice given the roadkill of LMS companies and projects that would come and go during the era.[5] Blackboard's survival in this competitive space insured the CSUDH campus a consistency of product from year-to-year and provided a steady platform from which programs like the SLP could begin to grow and thrive. But consistency in technology is a double-edged sword with innovation often suffering. Even in its earlier iterations, Blackboard users often complained about a lack of breakthrough innovation for the product ranging from its antiquated underlying code base, to concerns about long delay times fixing bugs, to worries about help desk response delays for issues big and small, to issues with how content could be more easily produced and disseminated (Eisen, 2009).

Through its intensive use of Blackboard, the SLP almost immediately recognized some of the core limitations of the product especially as newer cohorts entered the program. One of the most important issues was how content could be more easily managed across the template system that the program had developed earlier to address the multiplier effect. Even with the domain of change reduced to seven core master templates for the program, adjusting specific documents and other sorts of content changes threatened to absorb much of the limited time resources of the lean SLP staff.

Around 2005, a new concept was introduced to the internet: embeddable content. As web browsers became more sophisticated, content from one website could be easily embedded into another website utilizing a few nifty code tricks. One of the innovators in this movement was a company called Box. At the outset, Box was little more than an online storage repository for documents of many kinds including Word documents, PowerPoints and PDFs. But it soon implemented a radical way in which to share that content using "code snippets" like these which could be copied-and-pasted into another site like Blackboard:

```
<iframe src="https://app.box.com/embed/s/s60ok3o53fiasuafoayh1f6kny8q0faw" width="500" height="400" frameborder="0" allowfullscreen webkitallowfullscreen msallowfullscreen></iframe>6
```

FIGURE 4.2.   Screenshot from a course shell of what an embedded code snippet from Box looks like after it is saved in Blackboard. The arrow is pointing at the Gettysburg Address which is published on Box servers.

With the production of code snippets like these, which now could be readily created and then spawn-copied into Blackboard section course shells, the SLP staff could drive and focus content production through Box. The iframe embed option offered by Box was a neat and clean solution to content management for the SLP. As long as the iframe code snippet embedded in Blackboard course shells pointed back to Box, changes to particular documents could occur outside of the LMS altogether and solely within Box. Thus, if a sentence needed to be added to a document or a date needed to be changed after templates had been copied to course sections—and that included making changes even *after* the semester had started—it could occur singularly in Box and the change would be reflected to *all* courses shells which contained the iframe embedded snippet. See Figure 4.2.

In effect, not only had Box introduced a key time-saving technology for SLP production staff, but it also had created a disciplined workflow reducing unnecessary redundancies caused by exchanging documents through email—which had been the document exchange norm before the advent of Box into the staff's routines. This was a remarkably powerful innovation on three levels: First, it mitigated the work required to maintain up-to-date content. Content could even be changed after the semester had started thus eliminating the dreaded "multiple minute changes to multiple section shells" which preceded Box embeds. Second, by reducing the friction inherent in changing content across multiple section shells, the SLP team could ensure a new level of quality control for specific content items. And, third, staff members were freed from potential task-repetitive time-traps thus allowing them to develop more high-touch interactions with the SLP students—especially those unfamiliar with online learning routines. Blackboard's relatively paltry options of content development and control were quickly superseded by Box's sophistication, and the SLP took full advantage of it.

Best Practice 3: Cheap is Good, But Free is Better

It is hard to imagine a K–12 school or a university without an email system. There are very few, if any, schools and universities in the United States today which do not have email systems for instructors, staff and students. But this was not always the case. Early on, at the outset of the SLP around 2005, reliable campus email for faculty, staff and students at CSUDH was not a given. Instructor and staff email was hosted on an undersized campus Microsoft Exchange server which was slow and often in between multiple daily reboots. Meanwhile, the campus student email system, though more reliable than the instructor and staff system since it was supported by third-party off-campus company, was remarkably stingy in the resources it allocated to students: storage was limited to a measly 50-megabyte limit per each account which basically told students the campus was not taking student email very seriously. The painful truth for any instructors or programs like SLP at the time: very few CSUDH students checked their email because it was so inadequate for their needs.[7]

As shocking as it might seem to have a campus email system for instructors, staff and students that practically nobody mistook as ready for prime time, it really didn't seem to matter: for most instructors who taught at CSUDH at the time, campus email was a needless distraction that they wanted nothing to do with. But for the SLP, email communication was an essential lifeline for the program since easy-to-manage and regular direct contact with students would keep them on task and focused per program objectives and goals. As Toni Issa Lahera, the SLP Director, puts it:

> At that time, without a dependable email system for our students, we would have been in big trouble. We created our Blackboard course shells with email triggers so that students would be informed of timely interactions and assessments. It's different now: campus has its act together as far as email for [students, staff and faculty] is concerned. But when we started, we really had no choice but to turn to something like Gmail and other Google products. And, of course, there are all the other tools, too, like Google Docs and Hangouts which we still use.[8]

As an integrated practice, the SLP staff insisted that all students go to the web and get Gmail accounts as a prerequisite for participating in the program. The rationale was simple:

- Gmail was "free"[9] in the sense that there was no monetary cost, it was easy to setup (just get a Gmail account).
- It was easy to use and manipulate.
- It offered a host of possible tools including Google Docs and YouTube which Google purchased in 2006.
- It also offered pop-up conferencing with the Hangouts suite which was and remains a core communications tool for the SLP staff and students.

The SLP student Google accounts serve as a kind of Swiss army knife where students in the program can use tools like Google Docs to easily and consistently produce content which can then be shared in the LMS course shell per assignment requirements thus obviating yet another possible cost for students who might have been forced to purchase the Microsoft Office Suite which—at the time—was not available to CSUDH students at discounted prices.

Best Practice 4: A Cohort Bonds on its Stomach

A crucial aspect of the success of the SLP is the emphasis it places on social bonding as a component of social learning. There is an engaging informality which the SLP office design and ambiance conveys to visitors, but underneath there is an abiding sense of maximum efficiency—even when socializing. The SLP has developed a sophisticated summer program where cohort members meet face-to-face in a relaxed environment punctuated with lots of food, drink and play sessions. But to maximize time spent in this seemingly laid-back environment, the

SLP developed an online virtual boot camp which all new program students must navigate and pass through to advance to the face-to-face summer retreats. After completing the virtual boot camp, students are familiarized with all major tools, software and applications used in the SLP programs.

This includes training for Zoom. Zoom is a state-of-the-art conferencing tool that quickly outclassed anything offered previously by the competition—including Blackboard's own Collaborate suite. Of all the conferencing products tested by the SLP, Zoom proved to be the easiest to setup and learn which was critical for program staff who wished not to be overwhelmed with conferencing software support questions. Any number of other conferencing packages had notable shortcomings that made the SLP staff reluctant to engage and purchase. Moreover, it immediately became apparent that Zoom offered a rich conferencing environment that allowed the SLP to create social learning bonds which are inherently necessary for cohort development, and to do so in a virtual online environment. As Ruben Caputo, who leads the SLP online support effort, puts it:

> Many of our program's teachers and administrators work over 40 hours a week and we needed to find a way to quickly and effortlessly host class meetings during the week. We had previously tried various platforms prior to Zoom that we liked such as Fuze, Adobe Connect and even Google Hangouts. But there was always some little "gotcha" like convoluted registration or hard-to-install plugins. After extensive experimentation with these tools, we noticed we needed something more user friendly and engaging enough to hold class discussions.[10]

Immediately, it was clear that Zoom offered the program a superior graphics experience which made meetings more intensive and engaging. Caputo extolls: "We were ecstatic about the high definition audio-visual quality. This alone made the switch from the other platforms worth it for us."

Furthermore, Zoom's easy-to-use breakout session capabilities proved to be an instant success as it allowed quick group creation where fewer numbers of students could engage one another while the professor made virtual check-up rounds to see how things progressed. Caputo was especially impressed with the pop-up group breakout options within Zoom: "The instructor is able to have multiple small group meetings going at the same time which allows those students to work privately from other groups. It's turned out to be one of the more powerful tools we use inside of Zoom."

The tool was an especially welcomed complementary addition to the SLP array because it became the main way in which staff interacted with individuals or groups of students in live sessions. Once students setup Zoom and had become adept with its usage, they were able share their desktop screen with staff which is crucial in speeding up and troubleshooting issues early in the beginning of semesters—yet again another high-touch aspect of how the SLP bridges digital divides. Caputo says, because of its superior screen-sharing capabilities, Zoom is noticeably more efficient for staff when troubleshooting a student's potential technical issues.

> Being able to provide support remotely allows our students to do it in the convenience of their own home. This has relieved students of much worry and heartache that they are not, in terms of their own personal technology skills, up to the demands of the program. Once they learn how to use Zoom, there are very few issues we haven't been able to resolve quickly and successfully.[11]

Along those lines, the virtual boot camp concept also serves another purpose: it alerts staff almost immediately to incoming SLP students who may not be up to speed with requisite technical skills for the program. An alerted staff member is able directly triage the issue with a high-touch approach that often includes Zoom coaching and training. It is the most time-consuming, labor-intensive interaction between the SLP staff and students, but it quickly produces long-term dividends as those digitally challenged students—now trained and ready for all core technical aspects of the course shell—can more readily walk across the divide, as it were, for more productive and engaging interactions with other cohort members. And once up to speed with other cohort members, social peer learning begins to take over and staff can stand down on high-touch interactions. Program staff, readied for high-touch engagement at the beginning of each semester, learned quickly that it takes a village to ensure that no student gets left behind.

Once students, staff and instructors finally come together in these face-to-face social learning situations, lots of "in between time" create learning bonds with fellow companions sharing in a similar social space. That emphasis on building bonds within the cohort produces several beneficial effects of which these standout:

- *Student-centered learning* which shifts the burdens of high-touch learning to the students themselves and away from staff. This, in turn, also produces independent learning and critical reflection situations as group peer review create self-regulated cohort accountability.
- *Post-graduation networking* which produces numerous auxiliary benefits including information sharing and informal peer review ("4 Ways," 2015). This is a key requisite for leadership development as cohorts share their knowledge and experience beyond SLP.

Orientation get-togethers and formal summer camps are designed to create a culture of equity, hope and positivity. The efforts can be quite elaborate involving stay-overs at local hotels. But the goals of these face-to-face encounters are clear: to allow entering students to become part of program's vision for their learning and leadership. As director Issa Lahera puts it: "Everything we do during those two days is highly experiential and intended to touch the head and heart equally. Our goal is to shift thinking as they begin their leadership journey."

Best Practice 5: Life is a Carnival. Take Full Advantage of the Fact
That Humans Learn Better When They Laugh

Numerous studies have repeatedly concluded that humor in education can be especially beneficial in building community (Henderson, 2015; Stambor, 2006; Weimar, 2013). Within a cohort, humor reduces stress and creates cohesion among cohort participants. A Pew Research poll strongly suggests that people who watch humorous news shows like the *Daily Show* exhibited higher retention rates than did individuals who consumed their news from more traditional sources like the *New York Times* or *CNN* (Kohut, 2007). To that end, a key member of the SLP production staff is Fidel Garcia who heads the staff's video production unit. Central to his work is the seemingly off-centered way he interjects his brand of humor into simple how-to videos. Emphasizing the process, Garcia explains:

> The idea was to introduce funny scenarios that students could relate to so that we could get across an important but, ahem, otherwise boring stuff. The challenge for me is to portray humor in a way that gets everybody's attention but still conveys the weight of the idea. We realized right away that the best humor is incorporated into our media in small doses. The idea is to entertain as well as to inform. We do this in hopes of creating a relaxed environment for students. A pleasant mood will foster an open mind and a positive learning experience within our program. We take our work seriously, but not ourselves.[12]

The attention challenges are many with professional working students who must balance home live, work life in often precarious school sites with situations containing all the distractions inherent in the contemporary world. As Director Issa Lahera puts it, "we hired Fidel because right away we saw his quick wit could break-through in ways that more mundane producers miss. We're really lucky to have him."[13]

Best Practice 6: Brand Your Content to Create a Familiarity With It

When entering an SLP course shell, many observers have noted how sparse the design is. The SLP website design team has intentionally reduced menu choices down to the most fundamental level to drive participants quickly to core essentials of the site: *Announcements, Assignment Matrix, Course Materials* and *Tools*. The Assignment Matrix is the heartbeat document written specifically for a given course shell. It is an embedded code object from Box that staff have worked-over time and again to an exceedingly high level of quality control. In fact, like the Golden Gate bridge, the Assignment Matrix is always "being painted" with updates and changes taking place all within the controlled confines of Box—even, many times, after the semester has started. See Figure 4.3. The document which occupies this space in every course shell has been painstakingly vetted, reworked, carefully designed with color, shapes and a sense of time sequencing to produce a

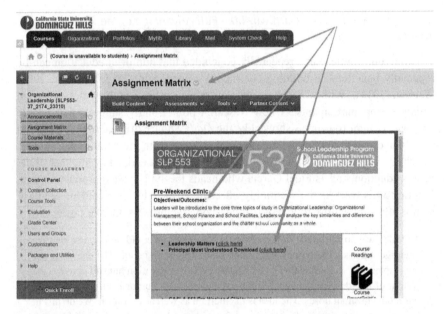

FIGURE 4.3. Screenshot of an Assignment Matrix in Blackboard. The Assignment Matrix embedded in Blackboard from Box is the result of painstaking conceptualization and many re-edits sometimes over a muti-year horizon. Also note the limited menu choices to the left where Announcements, Assignment Matrix, Course Materials and Tools are listed. Clean and neat.

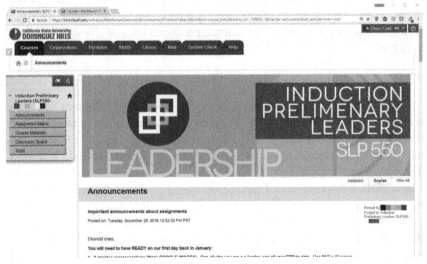

FIGURE 4.4. Detail of consistent color and icon usage inside a SLP course shell in Blackboard.

roadmap by which students will organize their time, study content materials and prepare for live virtual sessions via Zoom.

A notable visual consistency is immediately apparent when paging through an SLP course shell. The branded look and feel is the result of efforts by the team's graphic design director, Lupita Garcia. As Garcia explains, "What we're trying to do is to give our students a sense that the courses they take in the program are tied together and are related. Colors, shapes and content gather those strands and then differentiate them within the course shell. It's all done very intentionally. See Figure 4.4. Especially for visual learners, it provides a sense of visual clarity where words of explanation alone often fall short."[14]

The use of iconography lends itself to an immediately recognizable visual logic which reduces the anxiety that many students in the program may otherwise feel when entering a virtual classroom space for the first time. Icons serve as guide post and virtual landmarks that immediately help to familiarize students with their online learning environment. Garcia emphasizes the point: "We've done many usage feedback surveys with our students, and time and again they tell us how they really like how simple and straightforward our designs are and how that helps

FIGURE 4.5.   SLP brochure showing the "SLP Journey" which cohorts navigate through the program at the Tier I level. By Lupita Garcia showing icons and colors that inform design principles in SLP course shells.

them to quickly figure out what they need to get up and running successfully and quickly in their cohorts." She adds, "I'm a strong proponent of lean design for a lean team creating maximum efficiencies not only in aesthetic terms, but also in real-world navigation of otherwise potentially complicated websites and documents." For the SLP, investments in design have their payoff in fewer support questions thus freeing staff to concentrate on more high-touch interactions with students. See Figure 4.5.

Best Practice 7: To Build Leaders in Education It is Important to Remember: The Map Is Not the Journey, the Journey is Not The Destination, the Destination is Not The Map

Since Socrates and his famous seminars, educators have engaged in a seemingly eternal effort to brew the perfect pedagogic elixir which can deliver an intellectually balanced blend of teaching and learning. Despite its nascent promise to enhance education, the advent and establishment of the internet has, to the contrary, introduced an embedded array of digital divides that has not made the job of educators any easier. To the contrary, it has added newer dimensions and layers of potential inequity that can tank individual students and entire programs in its wake. If those dimensions are not thought-through in a considerate and deliberative manner, programs can collapse, and students can fail.

Informed by these potential traps and pitfalls, the SLP has instead created an equitable learning space where digital divides are acknowledged, recognized and addressed. The results have produced a program that is exceptional in so many ways, but perhaps its most important contribution has been its insistence on the individual empowerment of its students—many of whom are already well on their way to becoming school administrators with their own concerns involving educational technology and social equity at their own school sites. For those graduates of the program, instead of being the end of the journey, it is the beginning of a new one. As Director Issa Lahera underscores:

> Because we have students who enthusiastically use the technology and embed what they learn in their assignments, they take this skillset back to their schools. This process helps them to develop their own particular vision of how technology can work effectively in their own school sites—a vision that we help them to build from the ground floor. [15]

This is crucial because the crux of the SLP program is about developing leadership skills. Repeatedly, graduates of the program have emerged as school site leaders whether as master teachers or as enlightened administrators. Knowledge and experience gained from the program often catapults graduates into a higher order of leadership expertise for each her or his own educational pathway. As Issa Lahera puts it,

It is from here that they can return to their schools and help their students to prosper all the way up to the stars. That's not just syrupy pablum, either. We know of so many of our former students—now successfully situated in their careers that have used what they have learned with the SLP to create amazing school site learning spaces with the most minimum of resources. Our proudest moments are witnessing how they are transformed for the year they are with us and the potential that lies before them.[16]

CONCLUSIONS AND IMPLICATIONS

While finalizing thoughts and edits for this chapter, I had the opportunity to write the conclusion from the vantage point of a short stay in the Northeast of Brazil while visiting my in-laws celebrating the end-of-year holidays. While there, I was struck by how much time, effort and concern people the world over dedicate to education. Especially in countries such as Brazil which suffer from a narrower opportunity horizon compared to, let's say in the upper GDP income bracket countries like Denmark and Canada, education is seen as the key to social mobility however limited that might. In fact, in those places where job and career opportunities are narrowly restricted by class, race and gender structures, education is often seen as the *only legitimate possible* way out. Competition to get into limited slots in the nation's best educational institutions are often competitive beyond an outsider's comprehension. In Brazil, the educational coin of the realm is called the Vestibular.[17] It is a series of tests administered nationally for entry into the best universities across the nation. Because of the winner-take-all approach, it becomes a highly competitive endeavor. Annually, names of those who passed their Vestibular are listed on acceptance boards across the nation. The process is nerve wracking and creates a lottery-like fever among participants and family members. It is wrought with horror stories of multiple failed attempts to get into the country's limited set of state universities which are avenues for middle class entry and sustainability. For students caught up in this cycle, a failed entry exam is often the difference between a middle-class life and a life of socio-economic marginality.

In part to relieve this situation and to open newer educational opportunities, both public and private entities over the past few decades have promoted and invested in online learning as a potential learning alternative to the traditional Vestibular system which is clearly unable to keep up with the educational demands of the world's sixth largest economy. Even as Brazil's economy has stagnated in the past few years, the online education sector has continued to grow offering students alternatives to costly traditional face-to-face interactions. Already more than 25% of enrollments at Brazilian universities are for distance learning courses a number that is expected to grow to 40% as the economy begins to show signs of recovery (Teixeira, 2015).

Because of its increasing prominence as a key part of educational infrastructure of Brazil, *Educação a Distância* has garnered the increasing attention of many researchers who have asked many similar questions raised in this chapter such as what are the best practices for conducting online education, what are the best ways students can be prepared to engage in online learning, and is online education fulfilling the promises of complementing and even exceeding the current structure? The answer to those questions involves a robust and on-going debate made all-the-more intense because of austerity measure undertaken by the government which have adversely impacted educational budgets at all levels. Not unlike concerns of educational researchers echoed previously *and* currently in the United States, Brazilians worry about equitable distribution of resources, poor internet infrastructures, uneven access to computer power, gender/race/class distribution of resources and a host of other issues that researchers and specialists in the United States have traditionally called the "digital divide." One Brazilian study has articulated this concern:

> Success in student development is the primary goal of any teaching program. In this sense, the organization and management of distance learning courses require not only research on academic development, but also on how management processes and administrative support can contribute to student success (Martins, et al., 2017, p. 165).

In this sense, teaching the world over often boils down to efficacy and best practice within a set of shrinking budgetary resources. To do more with far less is the global mantra for educators everywhere. But with thoughtful leveraging of online educational resources, it *is* possible because it *has* to be possible. Necessity is the mother of these *best practice* inventions which this chapter has documented. The methods described here in this chapter are not unlike those emerging globally which attempt to facilitate newer modalities of teaching and learning in a dynamic and often fraught learning environment requiring acknowledgement of current digital divides and creative ways to deal with them. In that sense, the SLP shares much with the experiences of online educators in Brazil and the online teaching everywhere.

ENDNOTES

1. The good news is that with Moore's Law, chipsets for simple desktop computers have improved markedly so that school labs are not forced to upgrade to meet newer software demands. Also, most productivity software is moving to the cloud thus reducing resource demands on the desktop. It is far easier now to future-proof a lab for longer time periods than it was in the 1990s when lab-rot was an ever-present issue.
2. Both gender and accessibility issues were apparent from the very beginnings of the digital divide. What has changed is their central inclusion in the debate over equity.
3. Apologies to William Faulkner, Requiem for a Nun (1951).
4. Courses include: Organizational Leadership, Collaborative Leadership, Ethical Leadership, Post-Assess Prelim Leaders, Fieldwork, Professional Leadership: Pre-Assessment.

5. WebCT (1995, https://en.wikipedia.org/wiki/WebCT), CourseInfo (1997, later became Blackboard, https://en.wikipedia.org/wiki/CourseInfo), CourseNotes (2000), ePath (2000), CourseWork (2001), Moodle (2002, https://en.wikipedia.org/wiki/Moodle), Sakai (2004, https://en.wikipedia.org/wiki/Sakai_(software).

6. Code snippet from author's Box account pointing to copy of the Gettysburg Address.

7. These are my observations from my personal experience at CSU Dominguez Hills in the early aughts.

8. Interview with the author conducted October 13, 2017.

9. Google's idea of "free," of course, is different from how the SLP sees it. Google collects and markets its users' data and then sells that data to advertisers. To that extent, there is no "free lunch" even for consumers who otherwise do not pay a monetary fee for Google's services.

10. Interview with the author conducted October 14, 2017.

11. Interview with the author conducted October 14, 2017.

12. Interview with the author conducted October 18, 2017.

13. Interview with the author conducted October 14, 2017.

14. Interview with the author conducted October 19, 2017.

15. Interview with the author conducted October 13, 2017.

16. Interview with the author conducted October 13, 2017.

17. Because the testing is so standardized, critics complain that the Brazilian educational system is designed to teach for the test. Worse still, because they can afford tutors and other test preparation services, the rich are naturally advantaged over the poor (https://en.wikipedia.org/wiki/Vestibular_exam#Criticism).

REFERENCES

4 Ways Cohort Models Benefit Graduate Students. (2015). Retrieved from http://education.gsu.edu/4-ways-cohorts-models-benefit-graduate-students

Bodrick, J. (2016). Socioeconomic class issues in higher education. *NASPA—Student Affairs Administrators in Higher Education.* Retrieved from https://www.naspa.org/constituent-groups/posts/the-digital-divide.

Cohron, M. (2015). The continuing digital divide. *Serials Librarian, 69*(1), 77–86

Eisen, B. (2009). *A gripe session at Blackboard. Inside higher education.* Retrieved from https://www.insidehighered.com/news/2009/07/16/blackboard.

Faulkner, W. (1951). *Requiem for a Nun.* New York, NY: Random House.

Federal Communications Commission (2004). *E-rate and education (A history).* Retrieved from https://www.fcc.gov/general/e-rate-and-education-history.

Gazzaley, A., & Rosen, L. (2016). *The distracted mind: Ancient brains in a high-tech world.* Cambridge, MA: MIT Press

Henderson, S. (2015). Laughter and learning: Humor boosts retention. *Edutopia.* Retrieved from https://www.edutopia.org/blog/laughter-learning-humor-boosts-retention-sarah-henderson.

Herold, B. (2017). Poor students face digital divide in how teachers learn to use tech. *Education Week*. Retrieved from http://www.edweek.org/ew/articles/2017/06/14/poor-students-face-digital-divide-in-teacher-technology-training.html.

Kennedy, S. (2016). *The digital divide evolves. Information Today*, *33*(9), 8.

Kohut, A. (2007). Public knowledge of current affairs little changed by news and information revolutions. *Pew Research Center*. Retrieved from http://www.people-press.org/2007/04/15/public-knowledge-of-current-affairs-little-changed-by-news-and-information-revolutions.

Martins, R., Cruz, S., & Sahb, W. (2017). Relação entre a qualidade do trabalho da equipe multidisciplinar e o domínio conceitual sobre educação a distância. *Educação Unisinos*, *21*(2), 164–173.

Malburg, S. (2015). Understanding the basics of Title 1 Funds. *Bright hub education*. Retrieved from http://www.brighthubeducation.com/teaching-methods-tips/11105-basics-of-title-1-funds.

Mountain, A. (2017). Cross over the digital divide. *Principal*, *96* (4), 54–55.

Osborne, J. (2016). Alleviating the digital divide in the United States. *Childhood Education*, *92*(3), 254–256.

Parsad, B., & Jones, J. (2005). Internet access in U.S. public schools and classrooms: 1994–2003 (Elementary and secondary education). *Education Statistics Quarterly*, *7*(1–2), 8–11. Retrieved from http://nces.ed.gov/pubs2005/2005015.pdf

Paul, D. (2016). The Millennial morphing of the digital divide and its implications. *Journal of Negro Education*, *85*(4), 407–441.

Reynolds, R., & Chiu, M. M. (2016). Reducing digital divide effects. *Journal of the Association for Information Science & Technology*, *67*(8), 1822–1835.

Ritzhaupt, A., Liu, F., Dawson, K., & Barron, A. (2013). Differences in student information and communication technology literacy based on socio-economic status, ethnicity, and gender: Evidence of a digital divide in florida schools. *Journal of Research on Technology in Education*, *45*(4), 291–307.

Rogers, S. (2016). Bridging the 21st century digital divide. *TechTrends: Linking Research & Practice to Improve Learning, 60*(3), 197–199.

Stambor, Z. (2006). How laughing leads to learning. *American Psychological Association*, *37*(6), 62.

Teixeira, M.. (2015). Online education commerce in Brazil. *Tech In Brazil*. Retrieved from https://techinbrazil.com/online-education-commerce-in-brazil.

Weimar, M. (2013). Humor in the classroom: 40 years of research. *Faculty Focus*. Retrieved from https://www.facultyfocus.com/articles/effective-teaching-strategies/humor-in-the-classroom-40-years-of-research.

Winnick, M. (2016). *Putting a finger on our phone obsession*. Retrieved from https://blog.dscout.com/mobile-touches.

CHAPTER 5

WALKING THE PEDAGOGICAL LINE IN GRADUATE STUDIES

Obstacles and Opportunities Transitioning to Digital and e-Learning

Heather Rintoul and Duncan MacLellan

Over the last decade or so (Garrison, Anderson, & Archer, 2010) there has been an emphasis by universities to offer an increasing number of graduate courses online (Ferguson & Tryjankowski, 2009). It may now be an appropriate time to explore and investigate the perceived obstacles and opportunities to learning, given that a greater number of courses appear to be transitioning to an online format. In this chapter, we draw from an extensive literature of faculty perceptions (DeCosta, Berquist, & Holbeck, 2015; Topper, 2007) related to online graduate learning as well as our own pedagogical insights (Rintoul, 2016). Using the obstacles/opportunities paradigm we explore the following topics: virtual/physical presence, community of learners, inclusive democratic learning, innovative instruction, technology confidence, virtual evaluation and assessment, and cost effectiveness (Dobbins, 2009; Flaherty, 2010; Hauser, Paul, Bradley, & Jeffrey, 2012). There are numerous issues that are of on-going concern that will be queried for further research about online graduate education: ethical behavior, academic integrity (Starratt, 2005) and a shifting pedagogical paradigm related to globalization and leadership. These topics appear both intricate and multi-faceted and we therefore recommend mindful consideration of both aspects of the obstacle/opportunity paradigm that confronts faculty teaching online graduate courses.

Crossing the Bridge of the Digital Divide: A Walk with Global Leaders, pages 79–93.

INTRODUCTION

A Shifting Pedagogical Paradigm

As worldwide interest in social media and the internet continues to surge un-abated, a similarly strong demand for intricate electronic devices and resources of the virtual realm has exploded exponentially. Whatever one may wish to ac-quire today seems within reach by means of a computing device and access to information highway via the internet, even a college or university degree. With the intensification of undergraduate teaching in the virtual world, perhaps it was not unexpected that programs created for graduate level instruction would soon follow as the last bastion to succumb to an online modality (DeCosta, Berquist, & Holbeck, 2015). Advocates of undergraduate online teaching declared that the instructional paradigm of professor-centred instruction must be replaced with a learner-centred paradigm (Goodin, 2012). Unlike undergraduate education, how-ever, most graduate professors rarely lecture as the 'sage on the stage' and already regard the role of the instructor as one of facilitator to both support and encourage interactive dialogue and debate with the participant as the focus in the learning seminar (Barr & Tagg, 1995; Goodin, 2012). Consequently, there were those who assumed that the transition to online graduate learning might not be as difficult as initially imagined. The central question then was how to facilitate current learn-er-centred face-to-face practices with asynchronous online while still achieving successful pedagogical outcomes?

There are numerous matters for discussion around transitioning to the gradu-ate online paradigm. Even apparent opportunities, in the new graduate modality, may themselves expose unusual and unforeseen challenges online. Some obsta-cles have been relatively easy to solve, but others seem to defy resolution being more complex, unanticipated, and troubling. The first we review is pedagogy, which includes: virtual/physical presence, inclusive democratic learning, innova-tive instruction, academic integrity, and ethical behavior. From there we move to conceptualizing the community of learners including a discussion around student success, followed by technology confidence and finally, cost effectiveness. These matters are by no means disparate but rather present consequences and complica-tions that we argue interconnect and spill over one to the other, making resolution perhaps even more thought provoking and complex.

PEDAGOGY

Virtual/Physical Presence

With graduate students clamoring for a virtual learning experience (John-son, Stewart, & Bachman, 2015), universities were initially pleased to comply. Two positive aspects appeared to present themselves immediately: the university would see a reduction in faculty time and travel expenses to satellite campuses and at the same time satisfy student demand for virtual learning. Today's graduate student is often older (Brigham Young University, 2017), with a marriage or a life partner, family responsibilities, and full/part time employment. From the students' perspective then, virtual learning offers increased flexibility, an important consideration, given their hectic lifestyles (Johnson, Stewart, & Bachman, 2015). The opportunity to interact with other like-minded individuals from diverse cultures in various parts of the world has the potential to further enrich and broaden the learning experience for all participants.

Initially, virtual learning appeared to be an opportunity for universities, as it had (and still does) the very real capacity to reach all learners even from most remote corners of the globe, as long as they had electronic access, thereby increasing student base and institutional fiscal health (Dobbins, 2009). In the interim, however, graduate academic groupings for face-to-face instruction had quietly increased to 30 or even 35 to become the new norm. Graduate instruction online, though, has decreased, as universities have come to understand that for optimal learning, online supports no more than 20 in the participant community before becoming unwieldy and difficult to manage (Rich, 2015). The anticipation with virtual learning was/is that it offers exciting growth potential with innovative and dynamic pedagogical opportunities to access new modalities of knowledge and meaning-making (Kop, 2011).

Inclusive Democratic Learning

As colleges and universities strive to meet growing demands for educational outcomes that reflect 21st century priorities, this urgency has led to both opportunities and challenges for both faculty and students. Once viewed as "ivory towers" separated from the day-to-day economic, political, and social problems, postsecondary institutions have now become important sites to tackle these problems by engaging students as active citizens in democratic learning.

Inclusive education emerged out of a growing awareness and critiques of educational inequalities that had been historically embedded in the socioeconomic and cultural conditions of nations that were aimed at achieving an approach to education that was democratic and in the interest of the public good (Artiles & Kozleski, 2016). Lawy and Biesta (2006) note that when thinking about citizenship and inclusive education, there is often a tension between the traditional assumption of citizenship that focuses on citizenship-as-achievement and a growing emphasis on citizenship-as-practice, which provides a more robust entry point for understanding and supporting young people's citizenship learning.

Moving closer toward citizenship-as-practice has led to a host of innovative social experiments by universities to "tap" into student learning. Zeichner (2010) argues "…this shift toward more democratic and inclusive ways of working with schools and communities is necessary for colleges and universities to fulfill their mission in the education of teachers" (p. 89). The question then becomes one of answering which group(s) will benefit from inclusive education.

Innovative Instruction

Although the predominant format for communicating information in most university settings remains the "sage on the stage", increasingly educators are looking for creative ways to engage students to become active learners (Gilboy, Heinerichs, & Pazzaglia 2015). Recently, the flipped classroom is becoming a way for students to shed the traditional role of listening to teacher lecture and instead they read class materials and come to class prepared to engage in active learning strategies that may include debates and/or case studies. In this scenario, instructor and students are both immersed in the material (Gilboy, Heinerichs, & Pazzaglia 2015).

Rapid increases in internet access and advances in online technology enable us to reconsider how higher education courses are taught and students learn. A promising alternative to conventional lecture-based teaching is the flipped classroom instructional model (also known as the inverted classroom), which offers a framework for integrating emerging online learning technologies with active and collaborative learning (Galway, Corbett, Takaro, & Frank, 2014).

A study by Lee (2014) surmised that if professors teaching online courses demonstrated the following characteristics and behaviors: knowledge of the course, prompt reply, constructive and timely feedback on student works, these were important traits for students' learning opportunities. Nguyen, Barton, and Nguyen's (2015) examination of the costs and benefits of iPad technology as a learning tool, found that research is still at an early stage. While many higher education institutions promote the use of iPad technology, as integral to online learning, there is still a need to better integrate iPad technology within a holistic teaching and learning approach that focuses on student engagement and developing and delivering effective teaching services. Mobile technologies have the potential to become productive learning; however, more conclusive research is needed to draw stronger findings about the relationship between the increasing use of iPads and learning outcomes in graduate education courses.

Academic Integrity

As post-secondary educational institutions engage in more technologically driven forms of learning, the possibility for academic misconduct may increase. Therefore, many institutions are increasingly focusing resources on ways to educate students, faculty, and staff on the merits of academic integrity. Connected to this focus is the growing demand for educational institutions to "brand" their offerings as world class, innovate, cutting-edge, etc., which can be tainted by cases of academe impropriety. Macfarlane, Zhang, and Pun (2014) define academic integrity "…as the values, behavior and conduct of academics in all aspects of their practice" (p. 341).

Chertok, Barnes, and Gilleland (2014) noted,

...demands on students' time and resources may be increased as multi-tasking is an expectation in the digitally oriented reality, potentially playing a part in students' attitudes toward or participation in unethical behaviors. As such, institutions of higher learning, faculty, and administration need to take heed and address the impact of technology on academic Integrity to decrease the risk of violation and to promote a culture supporting academic integrity (p. 1328).

To avoid risks related to ethical issues posed by the online environment, it seems imperative that faculty to be aware of these potential risks. Faculty can then seek proper remedies and tools that promote strategies to reduce violations of academic integrity policies by student users (Chertok, Barnes, & Gilleland, 2014).

Often students do not fully understand the boundaries related to citation of intellectual property. Furthermore, respect for intellectual property is challenging to maintain when educators and administrators are unsure of the content that needs to be conveyed to students (Manly, Leonard, & Riemenschneider, 2014). Universities also need to ensure that their academic integrity policies are written in a manner that can be understood and interpreted correctly by students, faculty, and staff.

Ethical Behaviour

Both participants and the instructor, in an atmosphere of interdependence (Starratt, 2005), have an obligation to proceed in an ethical manner by 'attending' class fully prepared whether online or face-to-face. In traditional face-to-face learning graduate participants must, of necessity, make significant time and preparatory commitments to the learning experience: travelling to the location, reading/preparing academically, and dealing with all departure considerations for family such as meals and childcare. Online graduate learning does not have that embedded safeguard in that participants can 'attend' from a distance at unpredictable levels of preparation and contribution because no one, including the instructor, is physically present to verify participant readiness. Both face-to-face and online participants may be merely seeking course credit as their primary motivation, but with the instructor physically present in face-to-face teaching, strategic questioning can usually uncover non-preparation and remediate promptly.

With online attendance, it is significantly more challenging to discern breaches or interruptions in preparatory readiness. An online participant may not be pre-reading the required course material at all but rather merely responding to on-going conversational threads and with participants posting at different times, it becomes even more challenging to ferret out negligence. Participants '*at-a-distance*' may feel safer to neglect their academic preparation more than if they were attending with the instructor and their learning cohort collectively in one physical workspace, feeling they can escape detection because direct communication is unavailable when there is no one is physically present (Goodin, 2012). Although lack of commitment to appropriate academic standards online may eventually be discovered by the instructor, lapses in time wasted and non-preparedness robs others of any valuable interactions these neglectful individuals might have contributed to the learning community had they come online fully prepared (Hauser, Paul, Bradley, & Jeffrey, 2012). Online instructors may not feel comfortable adjudicating any considerations around failure of online preparation, fearing there may be personal, cultural or other mitigating circumstances and do not wish to be perceived as bullying participants (Clark, Werth, & Ahten, 2012; Park, Na, & Kim, 2014). Consequently, the indifference (perceived or otherwise) and poor groundwork of some may continue, resulting in less than an optimal learning experience for all members of the learning community.

Communicating online in an ethical manner offers both potential and opportunities but with some obvious concerns. Non-spontaneous online conversation presents an opportunity for each participant to take time to deliberately compose exactly what they wish to post that the rapid flow of face-to-face conversational interaction does not offer (Park, Na, & Kim, 2014). In online exchanges, however, all learners have the written word only to adjudicate intended meaning, which may, at times, be vague. The online reader is left to conceive and apply his/her own interpretation of the words, the meaning perhaps unintended, or may even regard the post as rude or abusive. In traditional face-to-face instruction, body language, facial expression, eye contact, as well as collective conversation and debate are available methods to illuminate, communicate, and clarify intended meaning in real time with the instructor present to facilitate and mediate when appropriate. Another potential obstacle to online conversational postings is the time it may take for participants to respond. Unlike in-person dynamic conversation, that time may stretch to minutes, hours, days or even longer, interrupting and disintegrating the conversational flow. In that interval, thoughts may have transformed and the juxtaposition of back and forth real time conversation to expand and build authentic understanding and learning through the immediacy of real time debate is just not available.

Group work is often a staple of online instruction giving learners opportunities to enhance their interpersonal skills in new and innovative ways, but if some students are not participating fully, for whatever reason, how can instructors ascertain who is doing/not doing what? What special strategies can instructors employ to scrutinize individual participation thoroughness in group work that does not interfere with the cooperative culture of the online community? Some studies have suggested that students should monitor or judge each other (Roberts & McInnerney, 2007). Asking learners to grade one another's participation or even to 'tattle' on those they deem not participating fully seems somewhat antithetical to the atmosphere of mutual support and learning so encouraged at the graduate level. Passing judgment on, or 'policing' each other without a personal relationship and understanding the myriad of mitigating circumstances or legitimate challenges participants might be facing seems a recipe for increasing anxiety and decreasing cooperation/learning. In face-to-face learning, participants may be more likely to feel an obligation to make others aware of an absence, as it will be immediately noted should they fail to attend. Individuals participating face-to-face tend to develop a more personal relationship with each other and the instructor, often sharing meals/shopping after seminars or staying to chat and share their personal lives at breaks and at the end of class. Although the social aspects of the new online learning cooperative have been much touted (Li, 2009), it may be that the 'silo' effect of an individual sitting alone '*somewhere*' in front of a computer results in a much more insular and individualistic learning experience than the more personal social interaction of face-to-face learning (Roseth, Saltarelli, & Glass, 2011).

The ability to write well has always been an important aspect of the graduate experience. If there is a significant difference between online writing quality in postings and the submitted paper at course end, how can faculty be confident that the registered student is actually the stated online participant and the creator of a submitted paper? Unfortunately, there are individuals who will attempt to behave in an unethical manner with the new modality (Kaufman, 2008) and deceitful practices may be even more challenging to uncover with online learning (Cabrera, 2013). Most instructors would like to believe that all students behave honorably, but there is a growing concern that the lack of checks and balances in online learning is just too tempting for those desiring to circumvent the system (Cabrera, 2013; Kaufman, 2008). In fairness to honest students, shouldn't instructors have confidence that students are receiving grades for work they have actually produced themselves? In this new online paradigm, it would appear that course leaders must construe online differently to authentically satisfy stated pedagogical goals (Wisneski, Ozogul, & Bichelmeyer, 2015), but how is this task to be accomplished? Most instructors have refined their instructional skills over time through personal experience, modelling and being mentored by professors they themselves have appreciated and regard as superior in pedagogical and instructional practices. Where are the mentors for navigating the pedagogical nuances of this new online instructional paradigm?

COMMUNITY OF LEARNERS

While the virtual learning paradigm offers new and unlimited access to resources and an increasingly diverse community of learners at any time and from any place, the one constant of the traditional graduate model has been of students and faculty corporally present in the same physical space. If the pedagogical aim is to support our students in an authentic way by modelling ethical purpose in an interpersonal environment (Leonard & Rintoul, 2010), how do we accomplish this task when we are not physically present in the seminar room? There are numerous differences between face-to-face and the current online modality which we will examine next.

Virtual learners are arguably more in-charge of their own learning as they navigate and interact in the online world. Instructors have long understood that graduate participants have varying aptitudes and needs. Self-directed learning (Goodin,

2012; Rosenberg, 2003) may offer a comfortable opportunity for some learners; but, without the scaffolding presence of the instructor to guide their experience, learners needing a more counselled involvement may not fare as well in meaning-making and knowledge-building (Helms, 2014). The online modality, with its access to a plethora of internet information, necessitates greater cognitive abilities, concentration, and meta-cognitive goals than face-to-face learning (Crampton, Raguse, & Cavanagh, 2012; Goodin, 2012). Content is rather flat information but gains capacity and dimension through interactive discussion and debate that challenges theories, beliefs, and practices thereby bringing substance to that which appears lacklustre, but do all students transitioning to online have the capacity for analysis and synthesis to determine relevant information on their own (Hauser, Paul, Bradley, & Jeffrey, 2012).

Online participants may learn from each other but how can instructors know what they are learning and perhaps more importantly who is doing the teaching? From a pedagogical perspective, how can online instructors successfully facilitate the learning experience and recognize when learners are in difficulty if we are not there in person to adjudicate? Instead of seeking help, some vulnerable members of the learning community may postpone seeking assistance hoping to 'catch up' on their own or, in some instances may not even realize that they are adrift at all!

Student Success

As demand for on-line courses continues to increase, both challenges and opportunities present themselves for consideration in terms of student success. Graham, Woodfield, and Harrison (2013), observe that when blended learning (BL), a combination of face-to-face and technological innovations, if not defined clearly and adopted strategically, institutions are not likely to really know the extent to which BL has been adopted on their campuses. Complicating this problem is ownership of intellectual property, which often becomes an issue with respect to BL implementation (Graham, Woodfield, & Harrison, 2013).

Challenged learners, those with some measure of disability, either major or minor, may find online learning more suited to their needs as they can choose when they want to be with their classmates more so than can the traditional face-to-face student. Learners who have auditory problems, have a linguistic difficulty (stuttering perhaps?) and an array of other challenges making them somewhat uncomfortable in the physical company of others, may do better online away from the pressures of face-to-face interactive conversation.

The number one myth according to a recent U.S. News and World Report article identified the misperception that online courses are easier, take less time to complete, and are not as rigorous as so-called '*brick and mortar*' courses (Murphy & Stewart, 2017). Murphy and Stewart (2017) point to recent research with undergraduate students demonstrating that successful course completion is lower in online class sections than in traditional face-to-face courses; although, meta-analyses have demonstrated that online environments are as instructionally effective as the traditional classroom. But, at least one study has indicated that unsuccessful online graduate course participants often achieve success by re-taking the same course face-to-face (Thompson, Miller, & Franz, 2013). Online learning, therefore, may not be for everyone.

TECHNOLOGY CONFIDENCE

With online instruction, the instructional leader has an opportunity to explore the synergy of innovative digital technology and course content in support of authentic learning (Moten, Fitterer, Brazier, Leonard, & Brown, 2013). With any new pedagogical aid, such as the new online model, there may be considerable apprehension about the technical skills necessary for comprehensive electronic teaching/learning. Instructors who are not digital natives may feel considerable anxiety as they struggle to understand the technical intricacies of the online paradigm. Similarly, these instructors may also be challenged merging content toward the best pedagogical outcome because of their less than stellar electronic capabilities (Goodin, 2012). Although faculty have experts to help them become proficient with the technical particularities of these generic instructional innovations, any personal modification and individuation around pedagogical soundness to align with program outcomes and practices is ultimately left to faculty to adjudicate and implement (Wisneski, Ozogul, & Bichelmeyer, 2015).

Course participants immediately anticipate that their instructor is their 'go-to' technical expert when electronic issues manifest. Troubles not resolved quickly and efficiently can have an immediate negative impact on the progression of learning within the online community. Even though faculty are aware that expert support is only a phone call or email away, technicians are available only during normal university working hours on business days, leaving online learning vulnerable should an electronic dilemma occur outside those hours (weekends and evenings). Technical delays are disruptive and may fracture learning, as the focus, of necessity, shifts to program accessibility issues rather than advancing discussion and learning around course content. As a result, instructors may feel increased anxiety and pressure at those times when their own technical lifelines are unreachable.

Online instruction presents an opportunity for students to participate that might be unable to attend a face-to-face seminar, but these learners tend to be very diverse regionally, demographically, and ethnically, as well as in their technical capabilities. It seems reasonable to comprehend that hardware availability can be sketchy in different parts of the world, with up-to-date equipment only an aspiration for some. Attempting to buttress the needs of these learners from afar presents its own unique obstacles for instructors and, until resolved, these challenges can negatively affect the progress of the learning community. Given the wide-ranging life circumstances of our participants (full/part time jobs and families) and unusual posting habits (e.g. early in the day, late at night, and from differing time zones), electronic glitches are more likely to occur outside university business hours when experienced technical assistance is unavailable. As a consequence, there is perhaps even more pressure for course instructors to be ready to problem-solve with superior technical expertise as electronic glitches can manifest at any time, even including weekends.

COST EFFECTIVENESS

The rapid rise in the use of desktop-based virtual reality technology in instruction is premised on the financial efficiency it offers in enhancing learners. Using desktop-based virtual reality instruction in educational settings involves not only monetary cost but also the efforts to train the educators to use them effectively. Therefore, it is critical that instructional designers make careful decisions in the design and development of instructional materials utilizing desktop-based virtual reality technologies (Nguyen, Barton, & Nguyen 2015). In a meta-analysis of online learning's effectiveness, Nguyen et al. (2015) concluded that strong evidence suggests online learning is at least as effective as traditional format but the evidence is not conclusive.

The goals of online learning, particularly at the undergraduate level, where classes are often larger and require additional administrative and financial resources are: to increase completion rate, speed-up time to degree attainment, reduce costs to postsecondary education, and offer more access to non-traditional students. Nguyen et al's (2015) meta-analysis of online learning's effectiveness, located strong evidence to suggest online learning is at least as effective as traditional format but the evidence is not conclusive. Online graduate programs (e.g. *BlackBoard* and *Elluminate*) are fairly cost-significant initially, however, and coupled with on-going annual subscriptions, update fees, and specialty maintenance technicians becoming the reality, the much-lauded fiscal benefit (Braddock, Mahony, & Taylor, 2006) has instead become a further drain on university coffers (Rich, 2015). Many course participants claim to prefer online instruction but a significant number of these have never experienced face-to-face graduate learning. Should the demands of students who have never experienced any graduate modality except online supersede philosophies, principles, and expertise of experienced faculty? Universities appear to be caught in the crosshairs between giving participants what *they say they want* and giving them what *they may need* to be successful. These two aspects are not necessarily the same nor mutually exclusive, but instructors, in most instances, are left to fall in line with university administration, which has decided that perceived financial viability trumps pedagogical soundness only to discover after the fact that the decision has fallen short of fiscal nirvana.

The fiscal cost of purchasing land and constructing teaching space, may be expensive for universities, regardless of location. For universities located in major urban centres, however, feeling the "space squeeze" as their enrollments expand, the competition with major property developers for much-sought after parcels of land can become very expensive. In the current economic environment, when provincial governments are not always able to support universities' expansionary needs, online learning can be a potential and viable solution. Providing virtual classrooms for graduate courses may help to remove pressure on universities to purchase costly land and engage in time-consuming building projects.

LEADERSHIP, GLOBALIZATION AND E-LEARNING

The internet has allowed information to become ubiquitous. In particular, Open Educational Resources (OER) and Massive Open Online Courses (MOOCs) have expanded in popularity, which reflect the global nature of these courses. These new approaches build on well-established traditions of distance learning and open universities (Haigh, 2014). E-learning and the Internet together have stimulated globalization at a pace heretofore unknown (Hongladarom, 2002). Western civilizations have assumed the leadership role embracing the new technology of e-learning through the Internet to other nations of the world. Today's learners belong to a connected generation. Their lives are interwoven and in this digital world, neither geographical location nor culture nor history matter, there is only the globalized on-line mixing bowl (Haigh, 2014).

As leaders of the new globalized e-learning mandate, post-secondary institutions must find the means to train instructors to deliver quality programming that is applicable worldwide to a wide variety of cultures. It would be short-sighted to assume that Western ideals are the best approaches to foster learning and knowledge-building. Instructors themselves may not have the necessary technical acumen or the knowledge to meet the requirements of a global education appropriate for all nations and cultures (Rintoul, 2016). The divide of technology is an obstacle because most conventional pedagogies are created to reproduce the cultural, social and economic hegemonies of Westernized educational systems (Haigh, 2014; Khiabany, 2003).

A Westernized model pervades the globalized culture of e-learning in education by focusing on what is suitable to teach that may disadvantage non-Western on-line learners (Haigh, 2014). To continue to support other cultures and 'difference' will require creative, inventive, and engaged re-thinking and action by e-learning leaders. The leadership of Western ideology has the responsibility to proceed with caution lest 'Westernization' through e-learning leads to the promotion of global cultural homogeneity and an eroding and subjugating of 'otherness' that is currently the desirable hallmark of each distinct society of the world.

Face-to-face graduate instruction has been honed over decades and centuries to become the excellent instructional and learning experience professors have crafted it to be, whereas the online capability is extremely new and, for the most part, largely untested (Rintoul, 2016). E-learning is part of an avalanche of social and technological change because it holds significant opportunities to create new pedagogical opportunities that bring people to virtual learning from different geographical regions (Haigh, 2014; Tuathail & McCormack, 1998). Both instructors and students are quite literally forging new ways of knowledge building and understanding within a new online modality that can be characterized as the 'far and away' physical isolation aspect of distance learning.

IDEAS FOR FUTURE RESEARCH

The ideas expressed in this work draw from a number of peer-reviewed sources that offered a variety of informed perspectives in relation to the challenges and opportunities that are before universities in the growing world of online education. In particular, we have focussed on the costs and benefits of online graduate education courses, which are pedagogically different from undergraduate courses. Some academic fields have made significant progress in tailoring graduate educational online courses to fit the educational needs of graduate students that could be beneficial to disciplines that are new to the world of online education.

Understanding the role technology can play in an online course is essential to creating an environment that uses technology positively to enhance teaching and learning opportunities. We also need to continue to research whether online courses are a cost-saving enterprise and the degree to which saving money should drive educational decisions. Some have advocated that online courses can increase course enrollments to a point that may be detrimental to creating opportunities for students to engage constructively. Finally, we need to understand whether students feel an online environment provides them and faculty teaching the course with a degree of intellectual engagement that is equal to or above a face-to-face classroom experience.

CONCLUSION

As we become more connected electronically within the university sector and beyond, pressure will continue to be placed on administrators and faculty members to offer online graduate courses. As universities continue to extend their reach into new areas of knowledge-making that may be conducive to online learning, it will become increasingly difficult to reverse this trend, for reasons that have been identified in this chapter.

Universities appear to be at an important crossroad with respect to online graduate courses, because there are pedagogical reasons that some courses should be offered, using an online format; however, we would argue there are graduate courses that benefit from a face-to-face format or some form of hybrid of online and face-to-face. The challenge for universities and faculty will be to walk the pedagogical line between online and face-to-face graduate courses while enhancing the learning experience for all.

REFERENCES

Artiles, A., & Kozleski, E. (2016) Inclusive education's promises and trajectories: Critical notes about future research on a venerable idea. *Education Policy Analysis Archives* *24*(43), 1–26.

Barr, R., & Tagg, J. (Dec. 1995). From teaching to learning: A new paradigm for undergraduate education. *Change 27*(6), 12–25.

Braddock, R., Mahony, P., & Taylor, P. (2006). Globalisation, commercialism, managerialism and internationalisation. *International Journal of Learning, 13*(8), 61–67.

Brigham Young University. (2017). *Graduate students by the numbers*. Retrieved from, https://registrar.byu.edu/graduation-numbers

Cabrera, D. (2013). Tips to reduce the impact of cheating in online assessment. *Best practices newsletter*. DeKalb, IL: Northern Illinois University. Faculty Development and Instructional Design Center.

Chertok, I., Barnes, E., & Gilleland, D. (2014). Academic integrity in the online learning environment for health science students. *Nurse Education Today, 34*, 1324–1329.

Clark, C., Werth, L., & Ahten, S. (2012 July/August). Cyberbullying and incivility in the online learning environment, Part I: Addressing faculty and student perceptions. *Nursing Educator, 37*(4), 150–156.

Crampton, A., Ragusa, A., & Cavanagh, H. (2012). Cross-discipline investigation of the relationship between academic performance and online resource access by distance education students. *Research in Learning Technology, 20*, 1–14. 14430—doi: 10.3402/rlt.v20i0/14430.

DeCosta, M., Berquist, E., & Holbeck, R. (2015, July). A desire for growth: Online full-time faculty's perception of evaluation processes. *The Journal of Educators Online, 13*(2), 73–102.

Dobbins, K. (2009). Feeding innovation with learning lunches: Contextualising academic innovation in higher education. *Journal of Further & Higher Education, 33*(4), 411–422.

Ferguson, J., & Tryjankowski, A. (2009 August). Online versus face to face learning: Looking at modes of instruction in Master's level courses. *Journal of Further and Higher Education, 33*(3), 219–228.

Flaherty, J. (2010, October/November). Bridging the digital divide: A non-technical approach to the use of new technology in post-secondary teaching and learning. *Academic Matters Journal of Higher Education*, 21–26.

Galway, L., Corbett. K., Takaro, T., & Frank, E. (2014). A novel integration of online and flipped classroom instructional models in public health higher education. *BMC Medical Education, 14*(181), 1–9.

Garrison, D., Anderson, T., & Archer, W. (2010). The first decade of the community of inquiry framework: A retrospective. *Internet and Higher Education, 13*(1/2), 5–9.

Gilboy, M., Heinerichs, S., & Pazzaglia, G. (2015). Enhancing student engagement using the flipped classroom. *Journal of Nutrition Education and Behavior, 47*(1)10–114.

Goodin, A. (2012, May 23). Online or in class: The shifting educational paradigm. *Distance and Online Learning*. Blog entry. Retrieved from https://evolllution.com/revenue-streams/distance_online_learning/online-or-in-class-the-shifting-educational-paradigm/

Graham, C., Woodfield, W., & Harrison, B. (2013) A framework for institutional adoption and implementation of blended learning in higher education. *Internet and Higher Education, 18*, 4–14.

Haigh, M. (2014). From internationalism to education for global citizenship: A multi-layered history. *Higher Education Quarterly, 68*(1), 6–27.

Hauser, R., Paul, R., Bradley, J. & Jeffrey, L. (2012). Computer self-efficacy, anxiety and learning in online versus face to face medium. *Journal of Information Technology Education, 11*, 141–154.

Helms, J. (20140. Comparing student performance in online and face-to-face delivery modalities. *Journal of Asynchronous Learning Networks, 18*(1), 147–160.

Hongladarom, S. (2002). The web of time and the dilemma of globalization. *The Information Society, 18*, 241–249.

Johnson, R., Stewart, C., & Bachman, C. (2015 August). What drives students to complete online courses? What drives faculty to teach online? Validating a measure of motivation orientation in university students and faculty. *Interactive Learning Environments, 23*(4), 528– 543.

Kaufman, H. (2008). Moral and ethical issues related to academic dishonesty on college campuses. *Journal of College and Character, 9*(5), 1–8. Retrieved from http://dx.doi.org/10.2202/1940-1639.1187

Khiabany, G. (2003). Globalization and the Internet: Myths and realities. *Trends in Communication, 11*(2), 137–153.

Kop, R. (2011). The challenges to connectivist learning on open online networks: Learning experiences during a massive open online course. *International Review of Research in Open and Distance Learning, 12*(3), 1–37.

Lawy, R., & Biesta, G. (2006). Citizenship-as-practice: The educational implications of an inclusive and relational understanding of citizenship. *British Journal of Educational Studies, 54*(1), 34–50.

Lee, J. (2014). An exploratory study of effective online learning: Assessing satisfaction levels of graduate students of mathematics education associated with human and design factors of an online course. *International Review of Research in Open and Distance Learning, 15*(7), 111–132.

Leonard, P., & Rintoul, H. (2010). An international collaboration: Examining graduate educational leadership in Louisiana and Ontario. In A. Normore (Ed.), *Advances in educational administration, 11. Global perspectives on educational leadership reform: The development and preparation of leaders of learning and learners of leadership* (pp. 301–321). United Kingdom: Emerald Group Publishing Limited.

Li, Q. (2009). Knowledge building in an online environment: A design-based research study. *Journal of Educational Technology Systems, 37*(2), 195—216

Macfarlene, B., Zhang, J., & Pun, A. (2014). Academic integrity: A review of the literature. *Studies in Higher Education, 39*(2), 33–358.

Manly, T., Leonard, L., & Riemenschneider, C. (2015). Academic integrity in the information age: Virtues of respect and responsibility. *Journal of Business Ethics, 127*, 579–590.

Moten, J., Fitterer, A., Brazier, E., Leonard, J., & Brown, A. (2013). Examining online college cyber cheating methods and preventative measures. *The Electronic Journal of E-Learning, 11*(2), 13– 146.

Murphy, C., & Stewart, J. (2017). On-campus students taking online courses: Factors associated with unsuccessful course completion. *Internet and Higher Education, 34*, 1–9

Nguyen, L., Barton, S., & Nguyen, L. (2015). iPads in higher education: Hype or hope. *British Journal of Educational Technology, 46*(1), 190–203.

Park, S., Na, E., & Kim, E. (2014 July). The relationship between online activities, netiquette and cyberbullying. *Children & Youth Services Review, 42*, 74–81.

Power, M., & Vaughan, N. (2010). Redesigning online learning for international graduate seminar delivery. *Journal of Distance Education, 24*(2), 19–38.

Rich, S. (2015, May 27). Personal email communication. Associate Vice President, (Academic). Ontario CA: Nipissing University.

Rintoul, H. (Feb. 2016). Chapter 27. The role of leadership and communication: (Re)-conceptualizing graduate instruction online. In A. Normore, L. Long, & M. Javidi (Eds.), *Handbook of research on effective communication, leadership, and conflict resolution* (pp. 515–530). Bethlehem, PA: ICI Books.

Roberts, T. & McInnerney, J. (2007). Seven problems with online and group learning and their solutions. *Technology & Society, 10*(4), 257–268.

Rosenberg, M. (2003 March). Redefining e-learning. *Performance Improvement, 42*(3), 38–41.

Roseth, C., Saltarelli, A., & Glass, C. (2011). Effects of face-to-face and computer-mediated constructive controversy on social interdependence, motivation and achievement. *Journal of Educational Psychology, 103*(4), 804–820.

Starratt, R. (2005). Cultivating the moral character of learning and teaching: A neglected dimensions of educational leadership. *School Leadership & Management, 25*(4), 39–411.

Thompson, N., Miller, N., & Franz, D. (2013). Comparing online to face-to-face learning experiences for nontraditional students. *Quarterly Review of Distance Education, 14*(4), 233–251.

Topper, A. (2007 December). Are they the same? Comparing the instructional quality of online and face-to-face graduate education courses. *Assessment & Evaluation in Higher Education, 32*(6), 681–691.

Tuathail, G., & McCormack, D. (1998). The technoliteracy challenge: Teaching globalization using the Internet. *Journal of Geography in Higher Education, 22*(3), 347–361.

University of Nebraska-Lincoln. (2016–2017). *Mentoring concepts for a dynamic learning community. Graduate Mentoring Guidebook.* University of Nebraska-Lincoln. Retrieved from, http://www.unl.edu/mentoring/who- graduate-students-are

Wisneski, J., Ozogul, G., & Bichelmeyer, B. (April, 2015). Does teaching presence transfer between MBA teaching environments? A comparative investigation of instructional design practices associated with teaching presence. *Internet & Higher Education, 25*, 18–27.

Zeichner, K. (2010). Rethinking the connections between campus courses and field experiences in college-and university-based teacher education. *Journal of Teacher Education, 61*(1–2), 8–99.

CHAPTER 6

A MODEL FOR ADDRESSING ADAPTIVE CHALLENGES BY MERGING IDEAS

How One Program Designed a Hacking Framework to Address Adaptive Challenges and Discovered the Ecotone

Kendall Zoller, Antonia Issa Lahera, and Julie K. Jhun

Digital shifts and innovations occur with a rapidity that create a never-ending learning curve. In the field of education, the slow adoption of technology in schools, along with the regular and rapid development of innovations, provides educators with the ultimate "adaptive challenge." Heifetz, Grashow, and Linsky (2009) define adaptive challenges as those "can only be addressed through changes in people's priorities, beliefs, habits, and loyalties." (p. 45). Their adaptive leadership framework serves as the bedrock of an emerging framework we have developed and are introducing in this chapter. Using and standing on their shoulders and of other great thinkers, we have designed a framework for working through adaptive challenges. We believe that acts of leadership create growth and change. In this chapter, we set the context for our framework as a leadership model. We then highlight the digital divide as the quintessential adaptive challenge for educators. The "Hacking Leadership" framework is introduced and the components within that framework are described. Finally, we introduce the "ecotone".

Digital shifts and innovations occur with a rapidity that create a never-ending learning curve. In the field of education, the slow adoption of technology in schools, along with the regular and rapid development of innovations, provides educators with the ultimate "adaptive challenge. Using Heifetz and Linsky's (2002) earlier work, and biology's "ecotone", we have designed a framework for working through adaptive challenges. The "ecotone" is the transition area between two biomes or distinct environments. We see it as a critical place where we go wittingly or unwittingly as challenges emerge. It is in this space that challenges can be grappled with and potential solutions (we refer to in this chapter as "hacks") are created. The ecotone is a place where we have permission to be uncomfortable and unsure. The levels of discomfort and uncertainty are within zones that are tolerable to us. The discomfort and uncertainty cannot be too high where people shutdown or too low where people ignore the discomfort and uncertainty.

All of the components of the Hacking Leadership model are necessary for navigating the ecotone. These components are the frame which we use to build our adaptive culture. Characteristics of an adaptive culture include embracing change, developing empathy, acting trustworthy, courage, and the willingness to function in an environment of ambiguity and uncertainty (Heifetz et al., 2009). We argue the use of the Hacking Leadership model provides a strategic process for addressing the digital divide as an adaptive challenge.

THE CONTEXT: THE DIGITAL DIVIDE AS AN ADAPTIVE CHALLENGE DEFINED

To understand the digital divide as an adaptive challenge, context is necessary. We look at the field of education as a sub-culture of our collective cultures. When looking at how technology is embedded in our everyday life and compare it to how technology is embedded in everyday schooling, we see a gap. The field of education has played an ever-widening game of "catch-up" with the technology industry for the last few decades. In almost every corner of the world, technological innovations, information and technology access have increased and transformed lives. This digital divide, however, has widened in the field of education. Hardware and internet access could not solve the digital divide. If we consider the number of technology initiatives for buying computers and tablets in schools nationally, it seems as if educational leaders and policy-makers have believed that if they put a computer in the hands of every student and teacher, the digital divide can be erased. However, the challenges surrounding access, knowledge and tools go deeper than the technical solution of merely providing computer hardware. This convergence of time, place and innovation is knocking at the schoolhouse door and brings a bundle of perils, pitfalls and gifts. How should educators respond? What are the real challenges?

With every new technology and mobile app update, we move from novice to expert and back to novice again. Every new iteration of what was once a familiar software program requires new skills and knowledge in order to navigate the new updated program.

Adaptive Leadership Briefly

Heifetz et al. (2009) offer two ways of looking at challenges. The first challenges identified are called "technical problems". Technical problems are "complex and critically important" (p. 45). They are also problems which may be resolved using existing knowledge, skills, and abilities. Adaptive challenges, however, are not solved with existing structures, knowledge or abilities. They require people to change their priorities, beliefs, habits, and loyalties. Moving toward resolving adaptive challenges requires people to "mobilize discovery, [while] shedding certain entrenched ways, tolerating losses, and generating the new capacity to thrive anew" (p .45). Solving adaptive challenges are not as simple as resolving technical problems.

In the real world, most problems have elements of both technical problems and adaptive challenges. The digital divide is a prime example of how technical problems are intertwined with and embedded in adaptive challenges. For example, a technical problem in the realm of the digital divide in education could involve installing the right computer equipment installed system-wide with sufficient technical support for the end user. However, even when these conditions are met, the digital divide remains if the remaining adaptive challenges are not met. Further, the adaptive challenges that may emerge might include how to encourage end users to adopt certain software usage practices, what the teacher-student relationship might be, how assessments could be monitored, and how the innovation could ultimately deliver positive educational results. Teachers and administrators must look at the digital environment in ways that generate new capacities while simultaneously shedding entrenched ways of teaching and learning. The generating of new capacities and shedding of entrenched ways are two factors indicating the digital divide as an adaptive challenge.

Overcoming adaptive challenges requires leadership. We believe leadership is a philosophy of relationships. To us, leadership, derived from Rost (1991) is a relationship of mutuality grounded in reciprocal care intending real change. Leadership is not mono, uni, or multi-directional. It lives in the relationships of those involved. Therefore, leading can emerge from anyone, at any time, and in any direction. Leading can be formal or informal. It can come from above or from within. It often happens simultaneously. Leading, we believe, is an autonomous act necessary to address adaptive challenges.

As we think about the digital divide, there is no definitive answer or solution, but there are incremental advances to be made. While the technical problem of providing technological hardware for everyone can resolve a practical issue, the persistent issue of persuading people from their entrenched ways to adopt technology and the potential benefits it brings will remain. What makes this an adaptive challenge is that we do not have the skills and abilities to fix it currently. We need to find people and knowledge and a place to do some work so that long held beliefs and practices can be safely tested and challenged.

"Hacking Leadership" Overview

"Hacking Leadership" is our iterative framework for implementation of the process of identifying issues and designing possible "hacks" (solutions). Figure 1 is a graphic representation of the model. Beginning with the initial challenge we split the model into two parts, the heart and mind. The heart is where our relationships live. It is where we mobilize the immunity-to-change model to develop empathy. The heart is the home of our values. This is the area where the application of communicative intelligence is helpful across our varied relationships. The mind on the other hand is the cognitive home for learning the content of adaptive leadership, adaptive schools, communicative intelligence, d-thinking and Dilts. It is home for intellectual acquisition of information to govern highly charged situations.

The heart and the mind are not necessarily in alignment at all times. We have choices on whether we choose to act through the heart or mind. These acts are deliberate. These acts can create the psychological permission and space to change. We must become skilled at heart and mind in order to grow individually and collectively. The heart and mind come together to name challenges and d-think for hacks (possible solutions). D-thinking, from the D-school at Stanford University, offers a program "where people use design to develop their own creative potential" (Stanford D. School, 2017).

Of course, other experiences may be orchestrated prior to and within this phase. Using protocols, values clarification, meetings, and processes where resistance emerges are some of these experiences. Out of the d-thinking phase come the hacks to be implemented on a small scale for testing. After testing, the results are analyzed to determine their potential system wide implementation. If a hack is deemed ready, it is implemented. If it needs modification, we modify. If the hack fails, it may be dropped or shelved. At the end of this process, if the issue is resolved, we have been successful.

The model looks simple (see Figure 6.1). It is also simultaneously complex and messy. In biology, the ecotone is defined as a transition area between two biomes. It is where two communities meet and integrate. We have adopted ecotone for our work. For use, the ecotone is the place between where we are, and where we are going. The ecotone gives us permission to be uncomfortable and unsure. The ecotone serves as the backdrop, the environment where shifts take place in who we are, and in our prioritized beliefs and values. The ecotonic environment can be real, cognitive, emotional, or spiritual. By naming the place where shifts in priorities of values, beliefs, and loyalties lie, we give ourselves psychological

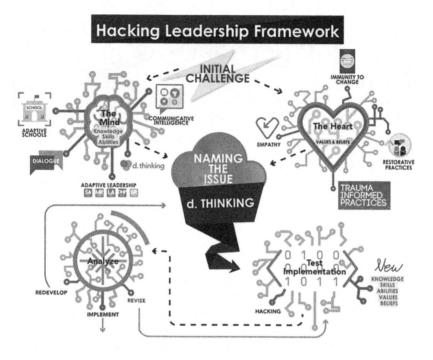

FIGURE 6.1.   Hacking leadership framework.

permission to be wrong, to change, and to develop. Naming the place "ecotone" lets us know we are entering a territory of uncertainty.

Let's review the model (See Figure 6.2) and define all of the elements more deeply:

1. Adaptive Leadership gives us a diagnostic tool set and a systematic way of thinking about challenges first through yourself and then with others (Heifetz & Linksy, 2002).

2. Adaptive Schools provide protocols for collaborating and building relationships. Relationships are the nexus of creativity and innovation (Garmston & Wellman, 2016). We use protocols from Adaptive Schools to guide the ways we talk and interact, thus impacting the relationships. Relationships are also influenced by our identities. We have many professional and personal identities. Each identity acts in unique ways in the same relationship. For instance, your identities may include supervisor, advisor, peer, mentor, content expert, and evaluator. In a meeting, how you talk and interact as a supervisor will look and sound differently when you talk as a content expert. By knowing our identities, we can shift our

FIGURE 6.2.   The Ecotone model connecting the current state to the possibilities when shifting identities, beliefs, values, capabilities, loyalties, and associations.

identities to shift the associations of our relationships an engage in ways we never engaged in before. These new engagements can serve as fertile ground for creativity and innovation.

3.   Immunity to Change gives us a process for looking at how we contribute to existing challenges ourselves and as groups (Kegan & Lahey, 2009).

4.   Communicative Intelligence (CI) gives us the tools to effectively communicate across our relationships as we move into and through the ecotone. CI helps establish levels of psychological permission to place adults in the 'zone of proximal learning' as they adjust their priorities of values and beliefs (Zoller, 2017).

5.   D-thinking is a process for creating environments where creativity thrives and innovations emerge. D-thinking is the process driving the creation of hacks. Hacks used to change the system resulting in new, previously un-thought of possibilities (Stanford D. School, 2017).

These are the five major elements of the hacking leadership framework. The framework is used to work through adaptive challenges. A significant theme within this framework is leadership. The subsequent section outlines and briefly explains the components of our leadership framework. By understanding the leadership framework, will foster the reader's understanding of how we applied it to the digital divide.

The Leadership Components Within

Our leadership components consist of five frames of focus (See Figure 6.3). These frames serve as the infrastructure to anchor the models named above. Within this framework lives adaptive leadership, communicative intelligence, adaptive schools, immunity to change, Dilts, and d. School thinking. We will use the following abbreviations to highlight the relationship between the foundational pieces and this framework:

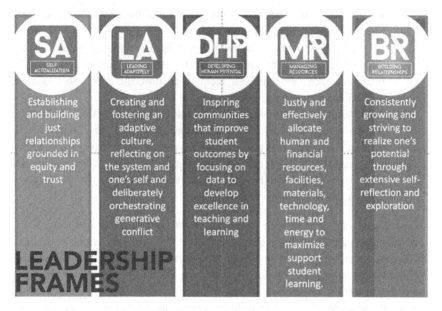

FIGURE 6.3. Leadership frames from CSUDH College of Education, School Leadership Program.

- AL- Adaptive leadership
- AS—Adaptive Schools
- CI—Communicative Intelligence
- ITC- Immunity to Change

1. Building Relationships (CI, AL, AS)- Consistently growing and striving to realize one's potential through extensive self-reflection and exploration. Building relationships with those we admire and respect. Building relationships with those who challenge our beliefs and values. Relationships where values among people vary and mismatch. One key to building relationships is to find common values and build from that foundation.

2. Leading Adaptively (Adaptive Leadership and Adaptive Schools)— Creating and fostering an adaptive culture, reflecting on the system and one's self and deliberately orchestrating conflict in ways that preserve and strengthen relationships while addressing the issues. Adaptive cultures embrace diversity of thought. It is in the diversity of thought that creativity thrives and innovation emerges.

3. Developing Human Potential (Communicative Intelligence, Immunity to Change, Dilts model) -. In all organizations, everyone must grow. Creat-

ing a culture where adults can learn in front of their peers is one characteristic of an adaptive culture.

4. Managing Resources (Adaptive Schools, Stanford d-thinking, and Adaptive Leadership)- Justly and effectively allocating human and financial resources, facilities, materials, technology, time and energy to maximize student support and learning.

5. Self- Actualization (Communicative Intelligence, Adaptive Leadership, Adaptive Schools, Dilts model, Stanford d-thinking, and Immunity To Change)- Establishing and building just relationships grounded in equity and trust. This is the development of a growth mindset where the impossible is possible where new paradigms emerge and our ways of looking at what is shifts to what can be.

The framework serves as the infrastructure of the content and processes students in our program experience. As we examine the digital divide, we believe the technical solution of providing every student a technology device is necessary but insufficient to achieve digital equality with all students. Shifts in practices and classroom environments must happen. More importantly, shifts in the values and beliefs about teaching, equality in the classroom, and universal access must happen. Our approach to digital quality has roots in technical fixes and adaptive shifts.

We stated earlier a focus on the work begins with a look at yourself. This graphic representation has its origins in Cognitive Coaching (Costa & Garmston,

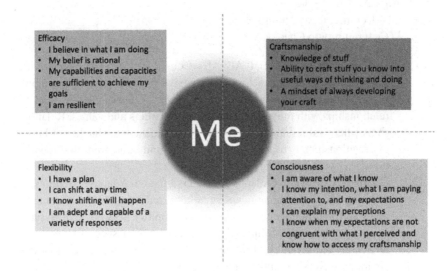

FIGURE 6.4. Adapted from Garmston and Wellman Adaptive Schools 5 Energy Sources.

2002). We use Figure 6.4 to develop personal consciousness about your capabilities, capacities, and choices. We believe we have choice. Choice in the actions we take. Choice in our responses to people and situations.

To be effective in the ecotone we need to be aware about what we know, intend, have capability to do, perceive, and most importantly when incongruence between our intention and perception surfaces. However, awareness is not enough as it would not be useful without having knowledge and skills. Knowledge and skills are what we call craftsmanship. Our drive for improvement and development is the catalyst for craftsmanship. We must have both awareness and knowledge and skills in order to grow. We continuously work to develop our knowledge and skills in a variety of contexts so we are prepared for the imagined and unimagined that can happen in the ecotone.

Efficacy is the flame driving our progress. Efficacy moves us when the implacable bears down to halt us. Efficacy cannot be held in your hand. It does live in your heart and mind. It is knowing that survival is the option in the ecotone. Flexibility is the *espirit de adaptivity*. Flexibility is what we have when things do not go as planned. That is what life is, a plan for the unplannable. With our knowledge and skills, we go forth with consciousness wrapped in a spirit of high efficacy knowing we will be adeptly reacting to the roils, fires, and precipices found in the ecotone. Flexibility is about thoughtfulness and strategic deployment of what one knows and can do.

Leading Adaptively

Using adaptive leadership is about "thinking first." This approach begins with identifying and diagnosing the issue as either a technical problem or an adaptive challenge. This includes understanding the history of the issue, the past attempts at fixing the issue, and the needs of the stakeholders. Too many times in the past solutions have been hurled at issues without sufficiently understanding them only to result in flurries of activity and action with no long-term positive result.

Adaptive leadership also requires deep levels of empathy for stakeholders. Engaging in dialogue as well as using protocols helps create an environment where empathy can surface. The adaptive leadership model moves from identification and diagnosis, through gaining empathy and into orchestrating conflict and acting political. This is where we rely on Garmston and Wellman's Adaptive Schools (2016), and Zoller's Communicative Intelligence (2017) as navigating this terrain requires skill in both individual and group interactions.

"Adaptive Schools" is a reservoir of protocols for interacting with stakeholders. The protocols offer structure and strategies for adults to work through challenges while honoring relationships. The "Assumptions Wall" is an example of a protocol we used. This protocol creates a visual representation of individuals within a group. It also creates an emotionally safe yet cognitively challenging environment for stakeholders to talk about the issue while honoring and respecting their relationships with others. This protocol is an example of separating the issue from the relationships. The assumptions wall is one of over one hundred protocols found in the Adaptive Schools work.

Adaptive Leadership and Adaptive Schools form the content foundation for the "Leading Adaptively" framework. The core of this framework is grounded always in communicative intelligence.

Building Relationships

Communicative intelligence (Zoller, 2017) identifies the verbal and nonverbal patterns of communication. Leaders will lead best when they can master communication. Communication is a cornerstone of the leadership framework because communication is often identified as the top skill necessary for leading and creating communities of trust (Heifetz et al., 2009; Kouzes & Posner, 2017).

The foundation of this frame is *The Choreography of Presenting* (Zoller & Landry, 2010) which outlines the seven essential abilities of effective presenters. These abilities include:

1. Establishing credibility—the notion that whoever is talking is worth listening to
2. Creating and maintaining rapport—nonverbal and verbal skills for building the social connections with people when communicating
3. Reading a group—the perceptional and cognitive skills used to know when people are engaged and thinking
4. Balancing task, process and group development—offers skills and strategies for balance how much of a message to deliver while simultaneously working to make and keep the group healthy
5. Listening to and Acknowledging participants—skills to use that lets a person know you are listening and understanding what they are saying
6. Responding appropriately—patterns to use to make the group right requires evaluating, synthesizing and delivering a congruent verbal and nonverbal message
7. Recovery with grace—An essential set of skills to use when your goals do not meet with reality when working with individuals and groups. It is about successfully navigating the unpredictable behavior of humans while striving for your communicative goal.

The seven essential abilities are important for navigating change because the manner in how we communicate defines our relationships. Often times how we communicate with those who we agree with looks and sounds quite differently than how we communicate with those who we disagree with. By learning to communicate more congruently and authentically the more likely people are to accept what you have to say and to honor the relationship (Zoller, 2017).

Developing Human Potential

Leading adaptively and building relationships lead to developing human potential. This dynamic component relies on elements of adaptive leadership, adaptive schools and communicative intelligence. From Adaptive Leadership (Heiftz et al., 2009) the concept of causing disequilibrium sufficient to exceed the threshold of learning must be considered. The eustress of learning is that level of stress high enough to cause learning yet not too high to cause a shutdown in those doing the work. Educators are very familiar with this concept as it mirrors the Vygotsky moment in children. Vygotsky asserted that the proximal zone of learning is defined by the cognitive state where something is almost understood based on prior knowledge and experience. It is a place where students have the capability and capacity for greatest learning (Vygotsky, 1978). It is not too abstract to exceed their mental models. Heifetz et al. (2009) call this phase 'giving the work back' (p. 241).

Giving the work back means pushing people to levels outside of their comfort zone to learn and build their capacity. Building human potential is about moving people from where they are comfortable and capable to where they experience discomfort and uncertainty. We refer to this zone drawing on the work of Vygotsky. Those in education are familiar with the zone of proximal learning. Imagine if we could create a zone of proximal learning for adults. This zone can be reached by giving work back—beyond existing capabilities. It is a zone with energy high enough to surpass the threshold of learning. As with learning, this cannot be a zone where the energy is too high so it exceeds the limit of tolerance (Vygotsky, 1978). It is a sweet spot to find. There is no single algorithm. It is something we must feel for. We believe the zone can be created in the ecotone.

Accomplishing success in the ecotone often requires the use of protocols to guide the adult experience. Protocols apply focus while at the same time honoring relationships. Several are utilized from Adaptive Schools as well as the Coalition of Essential Schools. Communicative skills are drawn from Communicative Intelligence (Zoller, 2017; Zoller & Landry, 2010) and include:

1. The systematic use of voice. When sending information use a voice credible sounding voice. A voice of narrow tonal modulation. When putting ideas on the table or asking questions that challenge expertise, use a voice with higher tonal modulation. A modulation we are familiar with when asking questions.

2. The pause. This is a pivotal skill. A skill that can influence attention, heart rate, breathing rates, and chemical balance. So often meetings are filled with words. Filled so much that the space for processing is permeated with sound. The pause offers time to think, to process, to ponder, to care for one's self.

3. Third point is the cornerstone used to separate the issue from the relationship. A third point is simply a chart, easel, text, screen or other object

with information on it that people look at when thinking. It is the place to 'visually display' the issue. When people look at the issue (3rd point) they create psychological space between the issue and the relationship. We believe the relationship is the source of understanding, innovation, and creativity. Placing the issue outside the relationship (visually as a third point) creates a positive culture of collaboration and innovation.

4.   Paraphrasing is a skill many people use and few people have conscious access to. What we mean is people often have an intuitive ability to paraphrase. Which is wonderful for talking with people you enjoy talking with. People who do not trigger you and cause you to downshift mentally into a state of fight of flight. Fewer people have the intuitive ability to paraphrase those who trigger them. Paraphrasing is a craft. To be strategic requires mastery of the craft, consciousness of intent and affect followed by flexibility. Flexibility to adapt in the moment for word searches, mirroring, and rapport. The heated conversations where values and beliefs are challenged. Where professional skills are questioned. Where not knowing or understanding can be construed as being ignorant or weak. We have found that when communicative skills are developed with competency, such as paraphrasing, there are far less relationship challenges.

One of the other significant outcomes of competency in communicative intelligence in the component of Developing Human Potential. This component forces us to consider how we develop our existing relationships. We call changes in our relationships, associations. Meaningful relationships change over time. Just as our relationships with our parents change from when we were young children, teenagers, thirty-somethings, and ultimately as we move into our 50s and 60s the relationship changes. The changes include shifts from dependence to independence, from seeking care to giving care, from relying on to relying for.

Those shifts in how we associate with each other are why we call them associations. An example of shifting and developing associations comes from a teacher who moved to curriculum coordinator. As a new teacher, he quickly developed a mentee to mentor relationship with several teachers. As the years passed that relationship shifted to a peer focus, one steeped in equality and mutual respect. After the promotion, the association of peer shifted to leadership. Once the teacher became a curriculum coordinator the relationships with teachers shifted to a political voice of representation. The representation was at the local board level, the state level, and even the national level. Each shift in association resulted in a shift in how they talked, what they talked about, and how innovations were developed.

To support the idea of shifting associations in existing relationships we introduced a model that illustrated the connections and interactions between identity, values and beliefs, capabilities, and behaviors (Dilts, 1990). Think of this model (Figure 6.5) as represented by the Russian dolls that fit within one another. The

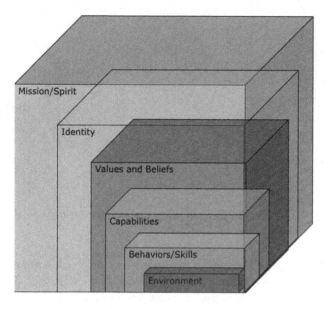

FIGURE 6.5. Dilts's model of change.

smallest being the environment. The largest being the mission and spirit. What the model infers is that whatever level you want to change, work from a larger level. When change happens at higher levels, the lower levels change on their own. We found when people focused on behaviors and skills long term, sustained change rarely happened. In many cases, most in fact, the resistance was implacable and the efforts were anemic in impact.

Shifting mindsets is one way to redirect their energy from behaviors to values and beliefs through identity. For educators, the values and beliefs around the sanctity of the content the purity of the subject are often untouchable. For instance, consider biology teachers who inquiry, laboratory exploration and the scientific processes. Using the Dilts model educators can shift to educational innovators of biology with a strengthening value of information gathering and processing through technology. As a result, what could occur is that technology that was considered to be a conduit for knowledge access and a tool for developing critical thinking might shift dramatically. By shifting behaviors and skills as teachers through technology change occurs.

Having worked through three of the leadership components let us look into the hacking phase of leadership. This is the phase where we apply what we have learned about ourselves and each other, as well as the adaptive leadership frame. This is the phase of creativity and innovation. In this phase, we introduce the d-thinking model from Stanford D-School.

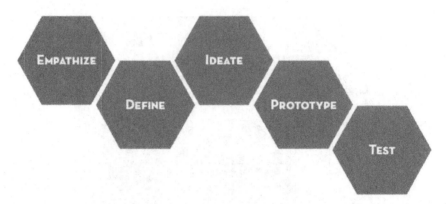

FIGURE 6.6.   Stanford d-thinking graphic representing the d-thinking process from empathy to testing.

The d-thinking model (Figure 6.6) begins with empathy. It relies on the effective use of communication skills from communicative intelligence. The purpose of d-thinking is first to understand challenges deeply and those it impacts deeply. From this understanding, in d. thinking problems can be identified. After identification, possible solutions are considered. We call these hacks. We define hacks as the innovations to test that will change the system. It is important to develop several hacks to test because a single solution may not work. One hack is chosen to test and data collected around the impact. This data includes stakeholders' feedback to the innovators. Modifications can be made at this stage for a second round of implementation. The implementation à feedback à modification à implementation cycle is how hacks are revised and refined.

As you think about hacks it is also important to set a context about change. We believe systems are immune to change. In fact, systems are designed to be stable. We also believe this to be true of ourselves. We develop neural pathways and reinforce them with repeating patterns. The pathways become habits and the habits become our patterns. Our own neurology is wired to be immune to change.

To address our own system of immunity we use *Immunity to Change* from Kegan and Lahey (2009). Using their work, people can work through a series of personal and group x-rays. The x-ray is a metaphor for the self-reflective protocol that looks at personal values and beliefs. It surfaces the values and beliefs that individuals feel proud of and that they perceive support their work. The immunity work also surfaces the values and beliefs which contribute to personal immunities around an issue. Consider how inserting technology into the culture of the school might impact each individual. Every person will have an individual set of values and beliefs, along with competing values around technology. It is these immunities that if left unexamined, make certain change never occurs. Values contributing to the immunity include, valuing competency, place of tradition, or curriculums believed to be nothing short of sacred. These values live beneath some deeply held assumptions. Assumptions are untested beliefs we have about something which we act upon as if they are facts and true. For instance, an assumption beneath the value of competency is that if peers at work see me struggling with content and pedagogy they will think less of me. The immunity to change protocols gives people viable ways of testing their assumptions so they can move beyond their immunity and change. It is with the addition of the *Immunity to Change* piece that we introduced the leadership frame of self-actualization.

Self-Actualization

Abraham Maslow's Hierarchy of Needs (1943) has been resurrected in education as a result of on-going research in marginalized student populations. In order to do powerful work, guided by a moral urgency and involving deep reflection and awareness of self and others, individuals needed to have a strong sense of self.

Maslow's premises were that if any part of the pyramid was compromised, deficits would emerge. Self-Actualization is a state where people are at their very best. Those who reach this state view the world with awe and wonder, are efficient in how they perceive reality, accept self and others, able to form deep relationships, able to express emotions and a life guided by values and goals. In order for adults to be able to understand and work with our framework, we believed they must be actualized first. Through exercises, simulations and case studies that was tied deeply to the work with Values, Communicative Intelligence and Immunity to Change we assisted all adults in moving towards actualization.

Putting all the work together thus far also included the addition of managing resources. This frame gave us permission to introduce the idea of systems. Leaders of schools not only define the culture of their school; they are also responsible for creating the systems necessary to manage the resources. We spent time looking at how to identify system gaps and system development

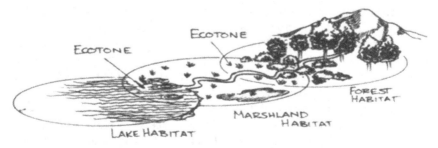

FIGURE 6.7. Adapted from Project Wild Ecotones are found between two or more communities.

Introducing the Ecotone as Part of the Hacking Framework

Every community shares a boundary with another community (See Figure 6.7). Some boundaries are distinct and well-defined. Others are less distinct and ill-defined. Oftentimes the boundaries are messy and offer challenges not found in each individual community. In biology, the ecotone is a place where adjacent ecosystems (communities and habitats) intersect. The intersection is an infiltration of both habitats. It is messy; the borders are ambiguous.

In order to get from one habitat to another you have to traverse the ecotone. The ecotone is rooted in two words. *Eco* which means place, and *tone* which means tension. The ecotone is a place of tension. Tension is necessary for growth. Drawing from biology, the simplest models illustrating this necessity is exercising. People generally exercise to become stronger. Exercise places stress on our muscles. It is a productive stress, a eustress, sufficient to grow new muscle, but not so much stress so as to damage muscle. Muscles will not grow without this resistance or stress and to that end, learning may not occur without an appropriate level of resistance or stress. The key is to create an ecotone with sufficient levels of eustress to produce learning and growing.

The richest ecotones are those with the greatest biodiversity. The greater the biodiversity the more likely those in the habitat will survive. Diversity is the nexus of adaptation. (Darwin, 1871) Survival in our use of the ecotone requires an adaptive potential. This means to survive in the ecotone you must be willing to shift what you do, why you do it, and how you do it.

In order to grow, learn, change, and move, we need a framework, a space and a way for doing this. From our experience, we know that ambiguity, uncertainty and conflict will emerge whenever there is change. (Heifetz et al., 2009). Instead of using the word conflict, we reframed this to the word "resistance". Resistance from a scientific perspective has an objective meaning in which is neither good nor bad. It is simply an influence on the flow of energy. When resistance is low, energy flows more freely. When resistance is high, the flow of energy is slowed (Heifetz et al., 2009).

All of the work with adaptive leadership, adaptive schools, communicative intelligence, Dilts, Immunity to Change, and d-Thinking gives people the 'what' and 'how' to move people from a place of stability and comfort and into a place of discomfort but safe enough to develop and grow and not too unstable as to disrupt beyond effectiveness. Navigating the ecotone is necessary to get to where you want to go. Think of the ecotone as a heroes' quest. It is full of conflict, resolution, and discovery. The ecotone is a place to overcome challenges. It is a place to practice resiliency to spring back from hardship, loss, and challenge.

Using the digital divide as an example, we can see the importance of the ecotone. As educators move to innovate and embed technology in the hands of students, they may discover that students have a greater proficiency and aptitude than the teachers teaching them. This stressful environment for teachers could become unproductive for teachers to learn and grow. The ecotone must be in a zone of productive disequilibrium for teachers to be successful enough to develop. Like the muscles we exercise, the stress must be tempered to levels of intensity sufficient to cause an environment for learning.

FINAL THOUGHTS

Adaptive challenges exist in our work and our lives. We believe survival of the fittest as an individual is a false idea. Survival of the most adaptive community is true. Strength is found in the variations of the community. That is the foundation of adaptivity. To endure within and though the ecotone, is about utilizing the adaptivity.

We believe any time one is working for change they are exercising leadership. We believe that the Hacking Framework is a way to solve the never-ending stream of adaptive challenges. The ecotone is a space to do the necessary work and move from where we are to where we want to be. Creating digital citizens at this time and place is an adaptive challenge that we can mitigate. Schools, and the educators who run them, have an imperative to do this work now in order to prepare able and responsible the global, digital citizens for tomorrow.

REFERENCES

Costa, A. L., & Garmston, R. J. (2002). *Cognitive coaching: A foundation for renaissance schools.* Norwood, MA: Christopher-Gordon.

Darwin, C. (1871). *The descent of man and selection in relation to sex.* New York: Appleton and Company.

Dilts, R. (1990). *Changing belief systems with NLP.* Cupertino, CA.

Garmston, R., & Wellman, B. (2016). *The adaptive school: A sourcebook for developing collaborative groups* (2nd ed.). Boulder, CO: Rowan & Littlefield.

Heifetz, R., Grashow, A., & Linsky, M. (2009). *The art and practice of adaptive leadership.* Boston, MA: Harvard Business Press.

Heifetz , R., & Linsky, M. (2002*). Leadership on the line*. Perseus Distributing. Boston, MA. Harvard Review Press.

Kegan, R., & Lahey, L. L. (2009). *Immunity to change: How to overcome it and unlock potential in yourself and your organization.* Boston, MA: Harvard Business Press.

Kouzes, J., & Posner, B. (2017). *The leadership challenge: How to make extraordinary things happen in organizations* (6th ed.). Hoboken, NJ: Jossey-Bass.

Maslow, A. H. (1943). A theory of human motivation. *Psychological Review, 50*(4), 370–396.

Rost, J. (1991). *Leadership for the twenty-first century.* Westport: CT: Praeger Publishers.

Stanford D. School. (2017). *Homepage.* Retrieved November 21, 2017, from, https://dschool.stanford.edu/

Vygotsky, L. (1978). *Mind and society.* Cambridge, MA: Harvard University Press.

Zoller, K. (2017). Using communicative intelligence to situate language in the context of restorative practices to create new associations. In A. H. Normore & A. I. Lahera

(Eds.), *Restorative practice meets social justice: Un-silencing the voices of "at-promise" student populations* (pp. 67–88). Charlotte, NC: Information Age Publishing.

Zoller, K. V., & Landry, C. (2010). *The choreography of presenting: The 7 essential abilities of effective presenters.* Thousand Oaks, CA: Corwin.

CHAPTER 7

EMERGING TECHNOLOGIES FOR LEARNING

Using Open Education Resources (OER)

Ruben Caputo

There has never been a more important time to access for all students to technology. As the gap between rich and poor increases, those not having access are doomed for life at the bottom. There is hope. Emerging innovations in education technologies have resulted in increased student engagement. The emergence of open education resources, or OER's, is an avenue that cannot only bridge the gap, but open new worlds of learning. OER's do not only increase student engagement. They are helping students and teachers bridge their learning relationship gaps by providing socially based tools and technologies. The innovative resources that have incorporated social computing technologies, will be showcased in this chapter. How these improve engagement will be seen through tools and pedagogies that are more personal, social and participatory. This chapter will reveal some of the key implications for practice, and conclude with an outline of the current challenges faced by educators.

WHAT ARE OPEN EDUCATION RESOURCES?

Open Education Resources are simply teaching, learning or research materials that are in the public domain or released with an intellectual property license that

Crossing the Bridge of the Digital Divide: A Walk with Global Leaders, pages 113–125.

allows for free use, adaptation and distribution (McGill, 2013). In an unprecedented move in 2001, the Massachusetts Institute of Technology (MIT) announced the release of almost all its courses on the Internet for free access. As a result of this move, several institutions followed prompting the number of institutions offering free or open courseware to skyrocket. The turn of events prompted the *United Nations Educational, Scientific and Cultural Organization* (*UNESCO*) to organize **the first Global OER Forum** in 2002 where the term Open Educational Resources (OER) was adopted. Later, with the partnership of Hewlett Foundation, UNESCO created a global OER Community Wiki in 2005, a move that was geared at sharing information and working collaboratively on issues surrounding the production and use of OERs.

Why Educators Should Use OERs?

OER's allow educators to adapt instructional materials to the individual needs of their students. This helps ensure that content and resources are up to date, relevant, and fit the unique needs of diverse student populations. Due to publishing timelines, traditional classroom materials like textbooks can often be out of date by the time they are used in the classroom. This does not even take into account the curriculum adoption cycles that exist in most districts, which result in content areas updating resources on a two-, three-, or four-year rotation due to budgetary constraints (Hale & Dunlap, 2010).

OER also guarantee that cost is not a barrier to accessing high quality, standards-aligned resources. Teachers can save significant time and effort related to resource development through the implementation of OER. Additionally, the open sharing of resources allows educators to collaborate across geographic, time, and space boundaries. OER showcases research to widest possible audience, shares best practice internationally and allows for peer review. The resources maximize the use and increases availability of educational materials.

Where to Look for OER's

So how do you find free, high-quality resources? When looking for OER, a good place to start is one of the repositories that houses a variety of tools for educators. OER Commons was created by the Institute for the Study of Knowledge Management in Education, and you can search a dynamic digital library of over 50,000 high-quality OER. Curriki hosts thousands of educator-vetted, openly licensed, online educational resources and allows for the creation of groups through which students and teachers can collaborate. As a joint effort of the U.S. Department of Education and the Department of Defense, the Learning Registry houses over 400,000 open resources for educational use. OpenEd describes itself as the "world's largest educational resource catalog" and has more than 250,000 OER aligned to standards for K–12 educators (Norris & Soloway, 2016).

What if you are looking for a resource and are unable to find it in one of the repositories? There are a number of ways you can search the internet to find what you are looking for without having to weed through everything that comes up with a traditional search. One way is to do a Google Advanced Search. In the options, you will find a field labeled "usage rights." Choosing the option "free to use, share, or modify" will allow you to locate OER. A Creative Commons Search allows you to access services provided by organizations that support OER. As a reminder, you should always verify that the work you find is under a Creative Commons license before you reuse, revise, or redistribute it.

EducationSuperHighway (n.d.) is an example of a nonprofit that is focused in upgrading internet access in every public-school classroom in America so that every student has the opportunity to take advantage of the promise of digital learning. The funding for such a reputable organization is supported by Bill and Melinda Gates Foundation and their mission is supported by America's leading CEOs. With efforts being provided to obtain affordable internet in the school districts this allows other organizations to continue to bring forth OER tools. There are several general sites that house a variety of OER, including MERLOT II, PBS LearningMedia, NCLOR (the North Carolina Learning Object Repository), OpenDOAR (the Directory of Open Access Repositories), and COOL 4 Ed (the California Open Online Library for Education). These sites provide access to tens of thousands of innovative, standards-aligned digital resources, student experiences, simulations, learning modules, assessments, and professional learning resources contributed by community members across the globe.

HOW OER'S ARE HELPING STUDENT ENGAGEMENT

OER has the promise of improving student engagement with course materials and can re-energize faculty engagement in course design and spark more dynamic approaches to teaching (Cortez, 2017). Redefining a lesson plan doesn't require the latest iPad app or tools designed with accessibility in mind. Instead, educators have brainstormed ways to incorporate music, video, clay modeling, trips outdoors, or tossing a ball around the class alongside the apps, and software they might typically use. The intention is to stimulate every type of learner, regardless of the individual support they might require.

There is evidence according to Cortez (2017) that suggests utilizing open content helps students better prepare for their courses and increases student engagement. The challenge is helping instructors curate and then package the open material in a way that is usable and part of a larger platform that also includes homework, assessment and other essential courseware elements. Engaging students can be difficult especially given how much they are faced with constant distractions. Furthermore, attention spans are getting shorter (Corte, 2017).

A high level of engagement should be a priority. It should not be surprising that the more students are engaged, the more they learn. It also makes it significantly easier to maintain your energy throughout the day. 21st century learners are growing up needing more than just the core curriculum to succeed in this digital age (Blair, 2012). Rather than viewing technology as a distraction, we should be taking advantage of it to increase student engagement. This according to Blair (2012) this makes the needs of the new 21st century learner a priority. To make sure that students are getting the most out of every lesson, the content should be presented in a way that the work has a clear meaning and immediate value to the students. Technology in the classroom allows students to gain a deeper understanding of topics that interest them, collaborate with each other, and direct their learning. The following is a list of seven key strategies for instructors to incorporate OER technology into your classroom to increase student engagement.

Submit Assignments as Blogs

- Blogs are short online articles (less than 3000 words) that are taking the world by storm. Blogs are a great resource for both the author and the reader: the author can constructively express their thoughts, and the reader learns something new.
- Best for students in grades 7–12 (Hall, 2013)
- Engage your students: Posting written assignments as blogs allow your students to showcase their work and also help each other out by posting comments. Posting the blogs publicly also means that you'll probably receive higher quality work.
- Add this to your classroom: Many free blogging services exist. Some of the more popular and user-friendly sites include *Medium*, *WordPress*, *Blogger*, and *Weebly*.

Submit Assignments as Podcasts or Videos

- Podcasts are digital audio files available for download to a computer or portable media player. Both podcasts and videos allow students to showcase their creativity and gain new skills.
- Best for students in grades 7–12 (Harris, 2017)
- Podcasts and videos are engaging ways to submit assignments that remove a lot of the presentation stress felt by students. They can create several versions and will not see the people who will be evaluating the assignment. You can also create podcasts and videos of your classes for your students to review after class or when preparing for a test. This also helps students who missed a class.
- Students can record videos by simply using the camera of a smartphone. Most smartphones will have a voice recording app, which can be used for creating podcasts. Have your students share videos through *Youtube* and podcasts through *Soundcloud*.

Work With a Classroom on the Other Side of the World: Global Learning.

- The Internet allows instant communication between anyone, at any time. Take advantage of your ability to teach your students about geography or history by talking to students from another country.
- Great for students in any grade (Svitak, 2010).
- Nothing is more engaging than a room full of students from another continent! Take pen pals to the next level by having your students chat and learn from someone from a completely different culture and background.
- *Skype* provides various methods of virtually adding people to your classroom, from interacting with other classrooms, to hosting a guest lecture, to virtual field trips.

Gamify Problem Solving

- Gamifying your classroom occurs anytime you bring competition or levels of achievement to a classroom exercise. Grades can be seen as a form of gamification. However, these don't necessarily fully reflect how much the student has learned or their work habits. Adding rewards to other aspects of the classroom allows you to focus on fundamental skills.
- Great for students in any grade (Aviles, 2017).
- Leveraging gaming mechanics can make learning more fun. Using your students' competitiveness also encourages them to work harder than they normally would.
- Work with your students to create badges or awards that they can receive for completing certain tasks, such as homework completion. Check out *Gradecraft* for some ideas. You can also introduce them to the concept of co-designing, allowing them to have a direct impact on what their favorite companies are making. For example, *My Starbucks Idea* accepts submissions of new Starbucks products and experience ideas, then allows users to vote on the ones that they like the best.

Create Infographics to Explain Complicated Topics

- Infographics are visual representations of information. They are designed to make information easy to understand very quickly. They are typically designed to draw attention to the most important information first.
- Great for students in any grade (Fanguy, 2017).
- Visuals are much more engaging than written text and convey a lot more information. Infographics take that to a whole new level. Infographics are typically used for displaying some sort of data; they could easily be used to compare population statistics in geography or show the history of the Internet.

- For an easy, user-friendly design interface, try *Piktochart*. Presentation software such as PowerPoint or Keynote can also be used to create simple infographics. More advanced design software such as *GIMP* or *Photoshop* are also great.

Record and Playback Reading

- Students do not always realize what their voice sounds like when they read. They may be making consistent pronunciation errors without even realizing it.
- Best for students in grades 1–3 or in foreign language classes (Redondo, 2018).
- Fluency and expression can be hard concepts for your students to grasp. When you record your students reading, they will be able to hear the differences in their voice and learn to recognize what it means to read with expression.
- Record each student reading using voice recording software on your computer, tablet, or smartphone. Then send this audio file to your students through email. If you want others in the class, or the student's parents to listen to the recordings, share them using Soundcloud.

Interactive Attendance

- Taking attendance by having students raise their hands can be time-consuming. This is important to know which students are present, but not necessarily the best use of your time in the morning.
- Best for students in grades 1–6 (Saxena, 2013).
- Make your students responsible for their attendance with an interactive whiteboard. This teaches them how to be accountable and gets them used to using an interactive whiteboard.
- Take a picture of every student during the first week of school and add them to a Notebook file. Each morning, have your students drag their photos into a "present" section when they arrive.

WHAT HAS OER'S DONE FOR OUR 21ST CENTURY LEARNERS?

In order to meet the challenges and encourage a real movement that will satisfy the need of the students, strategic planning is in order. Strategic planning requires analyzing what has been done so far. The planning needs to involve a vision and resulting implementation related to allocating budgets to expertise as well as educational realities existing in each country. What open educational resources have done to contribute to the bridging of this digital divide is provided informational content that is significant in the role of learning and teaching. It is helpful to consider learning resources by their levels of granularity and to focus on the degree to which information content is embedded within learning activity:

- Digital assets—normally a single file (e.g. an image, video or audio clip) sometimes called a 'raw media asset';
- Informational objects—a structured aggregation of digital assets, designed purely to present information;
- Learning objects—an aggregation of one or more digital assets which represents an educationally meaningful stand-alone unit;
- Learning activities—tasks involving interactions with information to attain a specific learning outcome;
- Learning design –structured sequence of information and activities to promote learning.

OER initiatives aspire to provide open access to high quality education resources on a global scale. From large institution-based or institution- supported initiatives to numerous small scale activities, the number of OER related programs and projects has been growing quickly within the past few years.

- In the United States resources from thousands of courses have been made available by university-based projects, such as MIT OpenCourseWare (n.d.) and Rice University's OpenStax project (n.d.).
- In China, materials from 750 courses have been made available by 222 university members of the China Open Resources for Education (CORE, n.d.) consortium.
- In Japan, resources from more than 400 courses have been made available by the 19 member universities of the Japanese Open Courseware (OCW) Consortium.
- In France, 800 educational resources from around 100 teaching units have been made available by the 11 member universities of the ParisTech OCW project (n.d.).
- In Ireland, universities received government funding to build open access institutional repositories and to develop a federated harvesting and discovery service via a national portal. It is intended that this collaboration will be expanded to embrace all Irish research institutions (Open Knowledge, n.d.).
- And in the UK, the Open University (n.d.) has released a range of its distance learning materials via the OpenLearn project, and over 80 UKOER projects have released many resources which are used to support teaching in institutions and across a range of subject areas. (Weller, Farrow, Pitt, & McAndrew, 2015)

With this being the current push to further promote OER initiatives to provide high quality education resources, the internet is a prerequisite source in order to obtain the OER goals. Currently, 99% of America's K–12 public schools and libraries are somehow connected to the web, in large part thanks to Federal Communications Commission's congressionally mandated E-Rate program which went into effect in 1998 (Ross, 2015). However, while that progress deserves merit, merely having some sort of Internet connection is an outdated standard. After all, that 99-percent statistic was achieved in 2006. Technology is integral to the modern learning experience, whether it's as simple as a basic Wi-Fi or as advanced as the artificially intelligent software that's replacing textbooks. In today's schools, having a dial-up connection is far from sufficient when measuring adaptation to modern times.

Digital Division

There is no escaping the fact there is a digital divide. The government has recognized this and is trying to develop e-learning strategies, particularly in vocational education and training, for groups like Indigenous students, people with a disability and the unemployed (Ross, 2015). There are many other groups, including rural and remote learners, isolated metropolitan learners and people with poor English literacy skills that also require assistance. Every group requires a different approach.

Steps are already being taken to address diverse needs but it is already easy to see where mistakes could be made. For example, a common goal is good, but a common strategy is not. Indigenous communities, for example, require a more nuanced approach than simple a one-size-fits-all policy. Indigenous people are no more homogeneous than anyone else. The same is true for the other target minority groups. Perhaps the cruelest part of the digital divide is the homework gap that students are facing. The "homework gap" is referred to as the barriers that students are facing on homework assignments without a reliable Internet source at home. This gap has widened as an increasing number of schools incorporate Internet-based learning into daily curriculum. Three-fourths of school districts, however are not doing anything about ensuring outside of school access to broadband (McLaughlin, 2017).

In 2009, the Federal Communication Commission's Broadband Task Force reported that approximately 70% of teachers assign homework requiring access to broadband. In addition, about 65% of students used the Internet at home to complete their homework, which could include submitting assignments, connecting with teachers and other students through group discussion boards, working on shared documents as part of a group project and doing online research for a school paper. Parents rely on the Internet as well to be fully informed on their child's academic performance, with many schools turning to online grading systems.

Less Workarounds, More Long-Term Solutions

These short access gains in winning over internet coverage bids to result in low-cost/savings initiatives, however valuable, are seen more as stop-gap measures until comprehensive strategies emerge to improve access to broadband in every community (Buckwell, Liberatore, Cruz, & Touchard, 2018). In 2015, the FCC announced plans to upgrade its Lifeline plan that provides telephone access to low-income households. This upgrade would extend support beyond voice service to allow participants to choose between applying the same support to either voice or broadband service. This simple change would both update the program and help bring more broadband to low-income households with school-aged children.

The FCC also modernized the E-rate program, implemented twenty years ago by Congress, which has been very successful in expanding Internet access to the nation's schools and libraries. In 2014, the program received an infusion of $1.5 billion to expand digital connectivity, a move strongly supported by the National Education Association.

All Our Digital Eggs in One Basket

Governments must be wary of being locked into one delivery mode. When the government made the commitment to give a "laptop for every child" it actually meant laptops for schools—a place children spend less and less of their learning time. Governments and universities must understand that online learning needs to omnipresent, preferably 24/7 and available on as many devices as possible, to meet the diverse geographical, cultural and resource needs of our students.

The Government has it right with the National Broadband Network: it provides the essential service of fast, widespread broadband access and leaves it to individual users and/or organizations to decide how best to exploit it for their own needs.

Flexible learning demands that governments and educational institutions at all levels engage directly with target communities, giving them the required resources, yet allowing the freedom them to use the resources to meet specific needs.

A Right to Online Learning

Online learning can address some forms of educational inequity, but it can also perpetuate others. We risk repeating the mistakes of the past and perpetuating educational inequity. Like access to private schooling, online learning comes with a cost barrier. Those who can afford smarter technology, better and faster connectivity and 24-hour access will benefit while those who can't, will not enjoy the same quality of access. We have to decide, as a society, whether high-quality online learning is a privilege or a fundamental right.

We are confronted with new Internet connection standards that currently need replacing and upgrading. Technology is integral to the modern learning experience, for any school to have a dial-up connection is far from sufficient when measuring adaptation to modern times. In 2014, the FCC approved additional $1.5 billion in funding for the E-Rate program, bringing a total annual budget to $3.9 billion. According to the administration, a typical school has about the same connection speed as the average American home but serves about 200 times as many users. Some schools even have to ration out Internet time to students. EducationSuperHighway, a nonprofit that evaluates school broadband speeds, says that situation leaves a lot of room for improvement.

OER CHALLENGES

There has been increased debate regarding the quality of OERs with some however arguing that these concerns are overstated. Either way, the concerns exist and must be addressed all together (Cox, 2016). Scholars suggest that the most obvious way of evaluating OER content is the same as for evaluating restrictively licensed content. The other aspect is that of use of *Peer Review* to validate quality, where a few Subject Matter Experts unanimously certify that the said content meets or exceeds acceptable criteria. This is seen to be one of the best quality indicators. Similarly, there is the *User Ratings* concept. Here, user ratings as used on some sites can be useful usually in the form of star ratings or numbers (Cox, 2016).

OER Instructors can also conduct self-evaluations of resources to ensure that the quality meets their standards. Additionally, the brand or reputation of the course developers or their institutions can be an important indicator of quality. Ideally, OER developers feel that while the above quality indicators can be used to gauge the quality of OERs, they should also be used in evaluating restrictively licensed content too (Yettick, Lloyd, Harwin, & Swanson, 2016).

Biggest Challenges That are Driving OER Adoption

In some places, the financial barrier to Education is worsening. Print textbooks make-up a $14-billion industry and have been hard to disrupt. For school districts, they create a major obstacle when education budgets have been tightening down. K-12 school districts, for instance, spend around 8 billion per year on textbooks. For colleges and universities, the majority of dropouts often cite high textbook costs as a major barrier to completion (Yettick, et al., 2016). An adoption to OER has lead data points to tremendous cost savings for K-12, higher Ed., and other learning environments. Teachers are already reporting significant changes in the classroom, as they've built their curriculum with OER while phasing out textbooks and other commercial teaching products (Yettick, et al., 2016).

Student lack access to important opportunities in U.S. school districts. In all types and sizes of districts, teachers and administrators still struggle with the conversation about education access. Whether it's a lack of technology resources to enrich the curriculum during school hours or the budget to provide tablets for every student to continue learning at home (particularly for families without mobile devices or computers), there are often low-income or other student groups who are getting left behind.

OER can help educators are looking to OER expand and improve the quality of classroom content. Since OER welcomes thousands of contributors around the world, it raises the standards for learning resources, enabling users to discover alternative ideas for teaching a subject and present relevant context and research. While schools may face limitations in certain subject areas, OER offers an enormous library of materials that can fill those gaps.

Teachers spend excessive amounts of time finding course materials. In American public schools, educators spend more than 53 hours per week on the job. Many school teachers are spending large amounts of time trying to find teaching materials.

OER enables teachers to draw from a very large (and constantly growing) collective of courses, readings, syllabi, videos, and quizzes. Teachers also have the opportunity to further personalize a course to better fit the diverse needs of their students. And after an instructor opens their own course materials with open educational resources, this becomes a transparent and collaborative way for other teachers to learn and share more frequently. Being a user and a contributor in the OER community means that learning resources can benefit from peer review to improve and evolve over time.

Many High Schoolers aren't college or career ready. The latest research from the National Assessment of Educational Progress (NAEP) (Kamenetz, 2016) has shown that most high school seniors aren't college or career ready. While the specifics are still not clear state to state, this trend has been connected with the increasing number of entering college freshmen who require remedial classes and other types of support, since they are unable to keep up with the first year's coursework. In the professional world, employers emphasize the importance of lifelong learning. Some education leaders are starting earlier in elementary school to build a college-going culture as well as build creative opportunities encouraging students to take charge of their own learning.

OER resolves this challenge by making available for free/low cost and accessible anytime online, creating this culture of lifelong learning. With more schools leaning toward OER, there is a change in education that puts the student at the center, empowering them to learn even after they have finished a course. With the growing concern over preparedness, OER has the potential to keep students engaged in their learning journey as they move forward to postsecondary education, a professional career, and life in general.

Effective leadership that focuses on all students learning is at the core of improved school districts. Leadership is committed, persistent, proactive, and distributed through the system. The two themes focus on all students learning and dynamic and distributed leadership are at the center in the conceptual model to illustrate their importance throughout the system as they connect and inform personnel, policy, programs, and practices in the district. A third theme—sustained improvement over time—indicates the forward and upward direction the district must take to have all students meet high expectations (Sue, Bylsma, Bergeson, & Heuschel, 2004). Having these leadership focuses will help close the digital divide gap we are currently facing. Being centered on principals that are caring about the accessibility of others such as low-economic communities will help bring forth more opportunities for the 21st century learner. With the right kind of people in leadership roles that affect student's abilities to learn the divide in which we see will be diminished. Improving student outcomes in achievement by improving or creating conditions that speak to this century learner will advocate sustainable improvement.

CONCLUSION

OERs are unlikely to replace traditional education; they nevertheless have the potential to obviate demographic, economic, and geographic educational boundaries and to promote life-long learning and personalized learning. The answer to equalizing access to knowledge and educational opportunities rests in OERs. School systems cannot alone solve this digital divide problem. Educators need to be a part of a community conversation and strategy of action that involves local elected officials, as well as the business and philanthropic community.

REFERENCES

Aviles, C. (2017). *Gamification in the 5th grade classroom: The conclusion.* Retrieved from, www.techedupteacher.com/gamification-in-the-5th-grade-classroom-the-conclusion/.

Blair, N. (2012). *National Association of Elementary School Principals: Serving all elementary and middle-level principals.* Retrieved from, https://www.naesp.org/principal-januaryfebruary-2012-technology/technology-integration-new-21st-century-learnerBroadband. (2015, December 08). Retrieved October 23, 2017, from https://www.fcc.gov/general/broadband

Broadband. (2015). Retrieved from, https://www.fcc.gov/general/broadband

Buckwell, M., Liberatore, F., Cruz, G., & Touchard, G. (2018). *Enabling rural coverage.* Retrieved from, https://www.gsma.com/mobilefordevelopment/wp-content/uploads/2018/02/Enabling_Rural_Coverage_English_February_2018.pdf

CORE. (n.d.). Retrieved from, https://www.core.org/

Cortez, M. B. (2017). *OER materials drive student engagement and savings, study finds.* Retrieved from https://edtechmagazine.com/higher/article/2017/09/oer-materials-drive-student-engagement-and-savings-study-finds

Cox, G. (2016). *The OER quality debate: explaining academics' attitudes about quality. The OER quality debate: explaining academics' attitudes about quality.* Retrieved from, conference.oeconsortium.org/2016/presentation/the-oer-quality-debate-explaining-academics-attitudes-about-quality/.

EducationSuperHighway. (n.d.). Retrieved October 23, 2017, from https://www.educationsuperhighway.org/

Fanguy, W. (2017). *Using infographics in the classroom: Our tips and advice.* Retrieved from Ppiktochart.com/blog/using-infographics-classroom/

Hale, J. A., & Dunlap, R. F. (2010). *An educational leader's guide to curriculum mapping: creating and sustaining collaborative cultures.* Thousand Oaks, CA: Corwin Press.

Hall, M. (2013, November 27). *Using blogging as a learning tool.* Retrieved October 23, 2017, from http://ii.library.jhu.edu/2013/11/27/using-blogging-as-a-learning-tool/

Harris, B. (Ed.). (2017). 8 *Terrific learning podcasts for students*. Retrieved October 23, 2017, from https://www.commonsense.org/education/blog/8-terrific-learning-pod-casts-for-students

Kamenetz, A. (2016). *Most high school seniors aren't college or career ready, says 'nation's report card'*. Retrieved from, https://www.npr.org/sections/ed/2016/04/27/475628214/most-high-school-seniors-arent-college-or-career-ready-says-nations-report-card

McLaughlin, C. (2017). *The homework gap: The 'cruelest part of the digital divide.'* Retrieved from, neatoday.org/2016/04/20/the-homework-gap/.

McGill, L. (Ed.). (2013, January 16). *What are open educational resources*. Retrieved October 23, 2017, from https://openeducationalresources.pbworks.com/w/page/24836860/What%20are%20Open%20Educational%20Resources

Norris, C., & Soloway, E. (2016). *So, when during the school day should teachers create curriculum?* Retrieved from, https://thejournal.com/articles/2016/04/05/school-day.aspx

Open Knowledge. (n.d.). Retrieved October 23, 2017, from https://openknowledge.ie/open-access-in-ireland/

OpenCourseWare, M. (n.d.). *MIT OpenCourseWare*. Retrieved October 22, 2017, from https://ocw.mit.edu/index.htm

OpenLearn from The Open University. (n.d.). Retrieved October 23, 2017, from http://www.open.edu/openlearn/

OpenStax CNX. (n.d.). Retrieved October 23, 2017, from https://cnx.org/

Paris Tech. (n.d.). Retrieved October 23, 2017, from http://www.oeconsortium.org/members/view/206/

Redondo, B. (2018).*11 Elementary language teaching apps you've gotta keep handy in the classroom*. Retrieved from, www.fluentu.com/blog/educator/language-teaching-apps/.

Ross, T. F. (2015). *The bandwidth gap: The states where school internet is slowest*. Retrieved from, https://www.theatlantic.com/education/archive/2015/03/the-schools-where-kids-cant-go-online/387589/

Saxena, S. (2013). *Top 10 characteristics of a 21st century classroom*. Retrieved from edtechreview.in/news/862-top-10-characteristics-of-a-21st-century-classroom.

Sue, S., Bylsma, P., Bergeson, T., & Heuschel, M. (2004). Characteristics of improved school districts. *Themes from Research*, 1–75.

Svitak, A. (2010). *5 ways classrooms can use video conferencing*. Retrieved from, http://mashable.com/2010/04/21/classroom-video-conferencing/#VRDdMRC6y8qj

Weller, M., Farrow, R., Pitt, B., & McAndrew, P. (2015). *The impact of OER on teaching and learning practice [Scholarly project]*. Retrieved from, https://oro.open.ac.uk/44963/1/227-1106-2-PB-3.pdf

Yettick , H., Lloyd, S., Harwin, A., Riemer, A., & Swanson, C. (2016). *Mindset in the classroom*. Retrieved from, https://www.edweek.org/media/ewrc_mindsetinthe-classroom_sept2016.pdf

CHAPTER 8

PARTNERING WITH TEACHERS TO BRIDGE DIGITAL DIVIDES

Doron Zinger, Jenell Krishnan, and Mark Warschauer

Bridging student digital divides is a longstanding goal of researchers, educators, and policymakers. Teaching plays a central role in student learning, so attending to teacher digital divides may also play an important role in bridging student digital divides. In this chapter, we present three cases of teachers who participated in professional development programs that partnered with the teachers. The programs were designed to improve student learning through technology. We detail the teachers' experiences and their own learning in the use of technology as an important lever for improving student learning. These three cases work to contextualize the complex nature of teacher learning in the use of instructional technologies. Across the three cases, we found the importance of meeting teachers where they are instructionally both in terms of technology and their subject area. Our findings also highlight the importance of attending to instructional approaches and pedagogy, not simply the use of technology in the classroom. Implications for teacher education and professional development are discussed.

Developing a digitally literate and technologically savvy populace are global priorities. School systems are charged with educating students in these digital literacies (Epstein, Nisbet, & Gillespie, 2011). Nonetheless, and despite increased availability of technology in schools throughout the world, a digital divide persists in schools (Incantalupo, Treagust & Koul, 2014; Warschauer, Cotten &

Crossing the Bridge of the Digital Divide: A Walk with Global Leaders, pages 127–144.

Ames, 2011). This divide often runs along socioeconomic status, as well as racial and ethnic lines (Pearce & Rice, 2013). One reason these divides persist is that they are not simply a matter of availability of technology (Tate, Warschauer, & Abedi, 2016). Whereas the proliferation of internet connected devices beginning in the early 2000s has bridged technological access gaps, gaps in skills and usage persist (van Dijk, 2017). In school, the persistent gap in student skill and use of digital technology also suggests that digital divides are part of a larger systemic gap and should not be viewed in isolation (Irvine, 2010).

When examining digital literacy in schools, a long history of disconnects exist between availability of technology, its use, and student learning (Cuban, Kirpatrick, & Peck, 2001; OECD 2016). Despite an extensive base of research reflecting their continued struggles (Cuban, Kirkpatrick, & Peck, 2001; Warschauer et al. 2011), school, districts, and educational systems continue to take "technocentrist" (Papert, 1990, p. 4) approaches to integrating technology into the classroom with poor results (Cuban, 2013; Gamze Isci & Besir Demir, 2015). Such approaches place emphasis on the technology itself rather than on instruction and student learning. A technocentrist may ask,

> Will technology have this or that effect? Will using word processors make children more creative writers? Or will it lead to a loss of handwriting skills? Will the computer increase interpersonal skills? Or will it lead to isolation of children from one another? (Papert, 1990, p. 4)

In contrast, more productive approaches to technology integration have been holistic and include the technology itself, technical support, as well as professional development (PD) for teachers who will be using the technology (Warschauer et al., 2011). These approaches highlight the importance and role of teachers and teaching in improving student digital literacy and bridging digital divides.

Thus, in this chapter we examine the role of PD in supporting teacher learning and bridging teacher digital divides, because we assert that this is an important step to bridging student digital divides. We present the cases of three teachers and their engagement in technology-focused PD. In doing so, we describe productive ways in which researchers, professional developers, and practitioners can address teacher digital divides to bridge student digital divides.

INSTRUCTIONAL TECHNOLOGY AND TEACHER PROFESSIONAL DEVELOPMENT

Technology rich teaching environments are critical for improving student digital literacy and bridging digital divides (Ainely, Enger, & Searle, 2008). Technology rich environments encompass: (a) the availability of technology; (b) the nature and types of curricula being taught; and (c) the nature of instruction in the classroom including teacher interaction with students and student to student interaction (Ainely, Enger, & Searle, 2008). Though greater access to technology helps, exceptional instruction and student learning can take place with limited technology (Ertmer, Ottenbreit-Leftwich, & York, 2007). As it relates to curriculum and instruction, student centered and constructivist approaches to instruction have shown positive student learning outcomes (Krishnan & Poleon, 2013; Rosen & Beck-Hill, 2012). In thinking about bridging teacher digital divides, we consider how teachers can design and develop technologically rich teaching environments. In the context of teacher learning, we also consider the limitations and constraints of PD. Specifically, PD programs typically cannot change the access to classroom technology but may be able to attend to those limitations in the design of teacher learning opportunities.

Professional development, in various forms, is a critical link to improving instructional quality and instruction with technology (Zinger, Tate, & Warschauer, 2017). Furthermore, PD that is responsive to teacher needs and centers on teachers has been shown to improve teacher knowledge of technology based instruction (Zinger, Naranjo, Gilbertson, Amador, & Warschauer, 2017). Indeed, comprehensive technological solutions that include involvement and support for teachers through PD have emerged as effective ways to address student digital divides by addressing teacher professional needs (Rosen & Beck-Hill, 2012, Warschauer et. al., 2011).

Productive ways to engage teachers in technology-based PD follow many of the same principles that generally guide effective PD. These include sustained engagement in the PD and a focus on effective instructional practices (Mouza, 2009; Rosen & Beck-Hill, 2012). An important technology-specific component is the need to address and support teachers' technological needs as well as their instructional and pedagogical needs (Polly & Orrill, 2012; Zinger et al., 2017). That is, PD should attend to both teachers' ability to use a technology-based resource tool, as well as how it can be used productively in instruction. Indeed, this is an ongoing challenge of PD programs, where teachers may know how to use a technology, but not how to use it to support student learning (Polly & Orrill, 2012). With this in mind, we present three case studies that highlight the importance of supporting teachers' understanding of how to teach with digital tools in order to bridge student digital divides.

CASE STUDIES

Given the complex nature of instructional technology and teacher education on its use, we present distinct cases of teachers learning how to teach with three, unique technologies: digital eBooks, a writing collaboration app, and a museum resources digital platform. These cases represent wide ranging instructional settings and teacher instructional contexts, as well as varying technologies and availability of technologies, as well as different PD approaches. All cases, however, represent instructional settings that are considered to be on the wrong side of the digital divide based on socioeconomic and demographic characteristics. These cases highlight both the possibilities and the limitations of supporting teachers as they learn how

TABLE 8.1. Case Study Teacher Demographics

Pseud-onym	Teacher Demo-graphics	Years Teaching	Teaching Setting	Student Demographics	Teaching Assignment	Available Technology	Experience With Tech	Digital Barrier and Divides
Anita	• Asian • Female	20	Large, urban, 7–8 school	28% English Learner 73% FRLP	English 8 Advanced English 8	Class set of Chromebooks, projector, document camera, teacher laptop	Piloted Chromebooks, 1 additional year, use of GAFE apps	Digital Literacies (i.e. Keyboard shortcuts: Ctrl + F)
Helen	• White • Female	5	Large, urban K–8 school	• 70% African American • 15% Multiracial • 14% White • 71% FRMP • 24% SPED	History Reading	6 Desktops, teacher ipad, projector and document camera	Extensive use of online resources such as YouTube	Availability of technology Connecting tech to learning goals
Frank	• White • Male	23	Small, rural, 6–12 school	• 93.2% White • 40% FRLP	English 8 Composition II (community college)	One-to-one iPad program, projector, teacher laptop	Piloted iPad program, 3 additional years, use of GAFE apps	Google Docs sharing functionality Interpreting program output

to use technology to meet student learning goals. Table 8.1 provides an overview of teacher demographics.

Case 1: Anita

Teacher Background and Teaching Context. Anita, a 20-year veteran English Language Arts teacher, and works at a large middle school in an urban school district in Southern California. She teaches 8th grade students, some of whom are English Learners (28%) and most of whom are economically disadvantaged (73%). For years, she has used the *Gradual Release of Responsibility Instructional Framework* (Fischer & Frey, 2013), providing focused and guided instruction using her projector or whiteboard and independent learning opportunities that rely heavily on paper-based methods. During the 2015–2016 school year, many of her practices changed.

Now, Anita teaches her 180 students each day in a newly renovated classroom that includes a class set of Chromebooks as a part of the district's grant-funded, one-to-one device program. Although she has vast knowledge of the instructional strategies that best serve her student population, Anita is new to the ways that digital tools can support learning in the information age. The district, too, is new to what productive teaching and learning with a one-to-one device program requires. Through some trial and error, the district realized that schools needed increased internet bandwidth, additional classroom internet points of access, tailored teacher support, and digital citizenship coaching for students.

Throughout the 2015–2106 year, Anita continued to collaborate with her colleagues and attend district-sponsored PD intended to support teachers' technical, pedagogical content knowledge. Because of her expert content and pedagogical knowledge, but still novice status as a technology user, the district's curricular team approached her with a unique opportunity to support a blossoming research-practitioner partnership (RPP).

PD and Teacher Learning Opportunities. Anita's district partnered with a local university to test the efficacy of a digital text formatting for improving students' reading comprehension. Visual Syntactic Text Formatting (VSTF) uses natural language processing to break sentences up at salient clause and phrase boundaries, fits each row of text into one or two eye fixations, and uses a cascading pattern to denote syntactic structure (Walker et al., 2007; Walker et al., 2005a). As one of this study's pilot teachers, Anita worked closely with the university's research team and the district's instructional team. In doing so, all members of this RPP hoped to learn about common stumbling blocks teachers new to the one-to-one Chromebook program were likely to face in general, and in using VSTF specifically. What the research team learned during the pilot year helped inform teacher PD opportunities intended to bridge the digital divide for 60 ELA teachers during the following larger implementation year.

Prior to using the class set of Chromebooks or the reformatted online materials in her own classroom, Anita attended two 6-hour professional development sessions held at the district's administrative office. These sessions were facilitated by university personnel and one teacher-consultant from an out-of-state district, but they were developed so that the pilot teachers could dialogue with presenters at any time. During this PD, Anita explored the reformatted eBook materials aligned with the school's current textbook, learned about the results of prior studies (Thomas et al., 2012; Walker et al., 2005b, Warschauer et al., 2012; Warschauer et al., 2013), and talked at length with the presenters about the instructional strategies that worked best for the teacher-consultant's students—students unlike Anita's from 1,000 miles away. These in-depth conversations helped to support the burgeoning RPP between the university and the district.

As Anita began using the reformatted eBook materials during reading instruction in her own classroom, she worked in partnership with the second author and research partner during 12 classroom visits throughout the year. The research partner was a former high school ELA teacher and experienced user of instructional technologies. Through more than 45 hours of classroom visits, a professional working relationship was established between this veteran teacher and the researcher. Whether during informal lunch conversations or semi-structured interviews, Anita had ongoing opportunities to reflect on her teaching practices as well as ways to meet her students' academic needs in a technology-enhanced classroom. This teacher-researcher partnership, bidirectional in structure, was successful due to its collaborative, exploratory nature. Both professionals worked in tandem to identify the best practices and instructional pitfalls that might occur during next year's technology-based literacy intervention and both shared the common goal of improving student learning. The extended, one-to-one nature of this professional development as well as its grounding in a purpose larger than Anita's own classroom are likely attributors to its level of productivity.

Through these classroom interactions, the researcher came to know which digital literacy skills were needed for close reading of activities using digital texts. During one interview, Anita openly shared her concerns about her students' motivation as it related to using the eBook materials during a recent classroom activity. She stated, "I think the ones who struggle are mostly the ones who have low motivation...[they] don't want to page through the paragraphs of the text." Given that VSTF breaks down long spans of texts into just a few words per line, even short stories become lengthy. Here's an example of how this RPP worked to address this identified challenge.

A few months into using VSTF, Anita stated that her students did not want to "page through the paragraphs." The researcher inferred that her students may not be using a keyboard shortcut (i.e. Control F) for identifying words or phrases when locating textual evidence, a digital literacy practice (Peres et al., 2004). The researcher responded by saying, "I wonder if using Control F on the Chromebook to find words might be a motivator, and give our students a quicker way to find what they're looking for?" Anita nodded with, "maybe that's going to help."

During the researcher's next classroom visit one week later, Anita's 8th grade ELA students were observed using the Control F keyboard shortcut, a navigational efficiency indicator, to identify textual evidence.

Through productive discussions with the researcher, the teacher targeted a current concern in the classroom (i.e. student frustration and motivation to locate evidence in digital text) and the researcher suggested one digital literacy skill that might alleviate it. But it was Anita's willingness to take a risk by changing the way she taught close reading via eBooks that was the first step in improving student digital literacy skills. Ultimately, this partnership was successful because it was built on curiosity, trust, and a focus on Anita's evolving professional needs related to technology use, as well as the intermediate goal of supporting a technology-based literacy intervention, and the ultimate goal of meeting all students' reading and writing needs.

From Learning to Action. During the broad implementation intervention year that followed, Anita continued to work with members of the university's research team through six additional classroom visits by the researcher and by attending monthly, 90-minute PD sessions. The Control F keyboard shortcut was presented during one of these PDs as an effective skill for locating textual evidence in digital texts. Her exploration of the affordances of VSTF, and her comfort with it, grew from what she came to know the previous year. Because of this comfort, Anita shared another effective strategy when incorporating VSTF into ELA instruction during the third monthly PD. To support her colleagues' understanding of her effective oral fluency lesson, she presented an overview to a room full of teacher-participants, including images of her students engaging in partner reading. This oral fluency lesson paralleled the one discussed by the out-of-state consultant during the introductory PD nearly one year prior. She reported that VSTF allowed her own students to read with more fluency. She recommended that partner reading (i.e. student pairs take turns reading VSTF text aloud supporting each other's pronunciation) be used to meet the foundational Common Core State Standard for reading fluency (CCSSI, 2010)—a skill that many of the district's English Learners were still developing. Other teacher-participants took up this practice once they had an opportunity to see it being used by their own district's 8th grade students.

The ongoing, one-to-one professional relationship between Anita and the university researcher helped transition her from an emerging instructional technology user to one who is confident enough to share her own technology-enhanced instruction with ELA teachers from all the intermediate schools in her district.

Case 2: Helen

Teacher Background and Teaching Context. Helen is a white, female, 6th grade teacher who teaches history and reading. She has been teaching for five years in a large urban school in the eastern region of the United States. The school she teaches in represents an urban, high-need school in many respects. The majority of students are African American and come from working class families, the school is considered low performing, and students need to pass through metal detectors to come to school each day. Helen's history class has a dozen students and all but one are African American. Student desks are set-up in a U-shape facing a table with a document camera and a projector in the middle of the room. Along the back of the room are six desktop computers with 17" monitors and there is a computer at the teacher's desk. Additionally, Helen uses a laptop and her personal iPad. In observing Helen, the research team found that she cares deeply about her students, often checking in on them and addressing issues related to their lives.

Helen uses technology regularly in her class, often showing videos from You-Tube, or gathering instructional resources from places like History.com. Helen notes the constraints that only having six computers in her classroom pose. She lamented for this limitation with, "I wish I had access to more computers so that the students could complete [their history activity] by themselves or in small groups instead of a whole group." Helen also raises the frequent double pronged challenge of finding time to plan and design, and coherent ways of integrating new technology that meet learning goals for students.

When Helen initially engaged in the PD program, she noted that "I already expressed my concern with 'time'. I still have concerns with the tool itself, or, how it fits into 21st Century learning... where learning is more about what you do with information/artifacts/tools rather than accessing it." This highlighted two challenges relating to planning time and pedagogy that PD designers and coaches would need to address through the PD series to meet Helen's learning needs. Both PD sessions and the coaching programs were deliberately set-up to be highly flexible and changed to meet the needs of individual teachers in their own instructional contexts.

PD and Teacher Learning Opportunities. Helen and two of her school site colleagues agreed to participate in a year-long PD program that focused on integrating digital museum resources in instruction. The PD program focused on the use of an online platform developed by a large national museum. The platform provides teachers with access to digital resources and tools to build collections of resources including photos, historical artifacts, and videos, as well as instructional tools to integrate collections of resources into instruction. The PD program was deliberately designed to engage groups of teachers from the same school site to promote collegial collaboration, which has been identified as a key component of successful PD (Garet & Porter, 2001). Unfortunately, due to a number of issues including the difficulty of finding substitute teachers, only Helen participated in the PD series.

The PD series included four day-long, face-to-face sessions running from fall to spring of the school year, three hour-long online PD sessions, and a coach that visited Helen's classroom five times and provided her with support via email and phone. The purpose of the PD series was to both develop teacher technological competence in using the online platform, as well as develop teacher instructional capacity to teach with the platform. PD time was provided for teachers to design activities and lessons. These efforts were further supported by instructional coaches that participated in the PD, and visited and supported individual participants in their own classrooms. As part of the program, teachers were asked to use the online platform for instruction, but there were no specific requirements of them. Some incorporated a single digital resource for activities and others used large, elaborate online collections of resources.

From the face-to-face PDs, Helen had opportunities to learn how the digital platform functioned, how the digital platform could be used instructionally, and how other teachers were using the platform. She also had the opportunity to plan and discuss her own use of the platform in her future instruction. During the first two PD meetings, significant time was dedicated to helping teachers learn how to navigate, search, and use various tools on the digital platform. As the PD sequence continued, more time was dedicated to demonstrating the type of instruction that could be designed and provided through use of the platform. Specifically, sessions were conducted to engage teachers in thinking and designing instruction that engaged students' historical thinking. Through all PD's, time for collaboration between teachers was provided in addition to the time designated for design.

Through coaching, Helen was provided with more targeted support. Her coach worked closely with her in designing and building collections of resources from digital platform and using them in instruction. Coaching support included the design of digital resource collections based on Helen's needs, co-creation of collections, sharing of instructional ideas, and feedback on implementation of the digital platform. As the school year went on, Helen's coach met with her to both plan and implement instruction, at times functioning as a teacher's aide, or second teacher in the classroom.

From Learning to Action. Helen first used the digital platform in January after attending two PDs and meeting with her coach. The platform was used over two days, where in the first day a video was used to introduce a topic to students and the second day an additional video and other materials were used. The materials were a combination of teacher-discovered resources including images and videos built on an existing collection of resources created by the coach. The use of video by Helen aligned with her previous use of digital resources in her instruction. She also used the collection in a primarily teacher centered way, projecting the materials to the entire class from her laptop. Part of the lesson included student discussion, but students did not get to engage directly with the digital platform or navigate it on their own. The collection and lesson revolved around their city and its history. Students were asked to write an essay about what it would have been like to live there in the past. Helen noted that the students enjoyed the lesson and activity, especially the aspects that connected to their own city.

The research team and her coach visited Helen in her classroom in February, the second time she used the digital platform and quickly became aware of the technological limitations in her classroom. Helen had set-up all her computers to the digital resource platform website and expressed her concerns about internet access and issues that she had run into the prior day. Due to the limitation of computers in the classroom, Helen had kids work in pairs or in triads for the day's activity. For this lesson, the resource collection designed by Helen was more student-centered than the first lesson observed. In this lesson, students navigated the digital platform and answered questions about the different resources aggregated in the collection. Similar to what the research team observed, Helen noted, that students were highly engaged with the activity. Central to this engagement was student agency and that the task allowed them to connect the digital resources they worked with to their own lives.

Through these two lessons we see how Helen shifted to a more student-centered lesson as she transitioned from the first to second time she used the digital platform. We saw her adapt to limited technology access by having students work in groups. We also saw her expand the use of the platform to become more student-centered as she became more comfortable with it, and as she saw students engaged with the platform. Indeed, in the subsequent year, other teachers adapted Helen's collection and she continued to use it. From a PD perspective, we saw the roles that face-to-face and coaching played in helping Helen contextualize learning for her students as seen from the first lesson. PD also supported her understanding of how to use the digital platform as well as how to teach with it, bridging the teacher's divide of utility of the platform and alignment with her instructional goals. Coaching was critical in supporting Helen, with the coach working to troubleshoot any technical problems that arose, create initial collections that met Helen's instructional needs, as well as encouraging Helen to engage with the digital platform.

Case 3: Frank

Teacher Background and Teaching Context. Frank, an English language arts teacher with 23 years of experience, works in the larger of two buildings that make up a small, rural public school district in Western New York. In the past, he has taught secondary ELA courses including English 9, English 12, electives, and an academic intervention course for struggling readers and writers. He also teaches composition courses at night at a local community college. Currently, Frank teaches 8th graders, some of whom are socioeconomically disadvantaged (40%) and almost exclusively Caucasian (93.2%). Unlike large districts who employ curriculum specialists, Frank is responsible for designing his own curriculum and making all decisions about the instructional strategies he will use each day, for each of his six classes. This means that he is free to explore unique methods and tailor instruction to his small classes of no more than 25, heterogeneously-grouped students.

Much like Anita, Frank's extensive teaching experience provided him with the pedagogical content knowledge to meet his learning objectives in a paper-based classroom (Koehler & Mishra, 2009). When the district's Technology Coordinator approached Frank about serving as one of the pilot teachers for the one-to-one iPad adoption, he was hesitant but agreed to teach ELA using the class set provided to him. Despite the outfitting of one-to-one mobile devices in his classroom, Frank was given little more than a tutorial on how to operate an iPad.

Frank's minimal guidance on how to maximize the utility of iPads as an instructional tool reflects this district's culture of classroom exploration as teacher professional development. But Frank and his colleague worked closely during the pilot year and shared the best practices and drawbacks to using iPads to teach English language arts. When Frank's colleague (the second author) left teaching to pursue a doctoral degree, she and Frank agreed to continue their professional working relationship that primarily focused on the effective use of instructional technology for reading and writing instruction.

PD and Teacher Learning Opportunities. Through attendance at weekly presentations, webinars, and research conferences, the second author learned about the latest instructional technologies. She often shared this information with Frank, and they discussed how various tools and instructional strategies might support Frank's current instructional goals. One writing collaboration program (Wang et al, 2015), an innovative new technology, held great promise for supporting collaborative learning in his classroom and the pair set forth on a classroom study of the effects of this tool on student collaborative writing—the first study of its kind (Krishnan et al., 2018). Frank, like many veteran teachers, understood the benefits and limitations of paper-based collaborative learning activities in the classroom (Cohen & Lotan, 2014). Ideally, students learn from one another, gain greater perspectives, and have a broader audience for whom to write. At its worst, one student is saddled with the responsibility of the activity while others make superficial contributions, causing resentment and frustration on the part of the students as well as the teacher. Despite these concerns, one of Frank's instructional goals was to design reading and writing activities that invited students to learn from, and with, each other. When his colleague explained that the collaboration program uses Google Docs' revision history to show how much each contributor to the document added (Wang, 2016; Wang et al, 2015; Yim & Warschauer, 2017), he quickly understood its power in meeting his goal—to encourage his 8th grade students to write together. Yet, Frank was unsure about asking his students to write collaboratively in Google Docs and was uncertain about whether this type of innovative learning activity was developmentally appropriate.

Frank nonetheless agreed to partner with the researcher on the study to determine if and how the collaborative program supports students' synchronous writing in Google Docs. Knowing however, that he would have limited on-site support when using this instructional tool, he had reservations. Despite this, the pair developed two argumentative assignments to be used in the study—one to be written by his 8th graders in triads and one to be written independently. Above all, the project was intended to meet Frank's instructional goals, aligning with the Common Core State Standards for writing which asks students to write arguments to support claims with clear reasons and relevant evidence (CCSSI, 2017). In an email sent to his research partner, Frank asked, "How do you interpret the program's data? How are we going to make sure that students only use in-class time to write in their Google Docs?" These questions helped his former colleague bridge the gap in Frank's understanding of the intervention tool and Google Docs' permissions. She created tutorial videos intended to support a greater understanding of these two important study elements. The study began once Frank and the researcher were comfortable with Frank's understanding of the tool and how to use it during instruction.

During the students' first day of writing, Frank asked them to use the collaborative program to see how they were contributing to their co-authored essays, taking time to highlight different collaboration patterns in different student groups and their implications for participation. After students left for the day, Frank contacted his research partner with an update. He shared the details of his unique and highly innovative classroom experience that only a former teacher could empathize with, and perhaps more importantly, help to troubleshoot. Through open dialogue, the pair negotiated ways to better support his students' online writing without contaminating their study's design. Frank also shared that he and his students were still unsure of how to interpret the primary data output for the program. Based on their initial use of the tool, the pair decided to make adjustments in the students' use of the tool by drawing their attention to a more easily interpretable numeric output table. In addition to troubleshooting solutions to classroom issues throughout this project, Frank shared his perceptions of the tool in supporting 8th graders' online, collaborative writing and how he thought it might be better suited to support the community college students he taught during his night classes.

Ultimately, this partnership and classroom study was relatively successful. Frank continued his professional development in the areas of data-driven writing instruction, classroom study design, and the use of instructional technologies for collaborative writing. Through this long-standing working relationship, the focus remained anchored in Frank's instructional goals. His willingness to share what he need further clarification on helped maintain the integrity of the study design and supported his instructional success.

From Learning to Action. Prior to the start of the following school year, Frank reached out to his research partner with his plans for using the collaboration program in the near future. He shared that he would be using it in his community college-level composition course, but did not report plans to use it with his 8th graders. He also shared his perception of 8th graders as being less suitable for group writing in general. This appears counter to the results of their classroom study, which found positive effects of group writing on students' subsequent, independent writing. Even after further dialogue about the positive study results, Frank's plans for using collaborative writing activities at the 8th grade level did not change, for this school year. However, Frank's use of this tool with his college students was later explained as his way of "getting it right." He shared that his college-level learners were better equipped to deal with the interpersonal complexities of online collaborative writing and would be more apt to succeed while he further explored online writing pedagogies. Much like the other two cases, Frank's focus was on student success and his adoption and adaption of tools was highly contextualized.

CONCLUSION

Our goal in presenting these case studies is to provide insight into challenges as well as productive approaches to bridging teacher digital divides with the intent of bridging student digital divides. We present three considerations that professional developers and researchers may find useful in working with teachers to address this objective. First, all three of our cases reflect a partnership approach in which researchers and PD developers work together with teachers—an approach that contrasts with much of the PD teachers experience. Because these partnerships were able to focus on teacher needs, solutions were no longer a top-down effort to affect change within the classroom, a problem often observed in one-size-fits-all PD (Darling-Hammond, 2017). Second, in all three cases, the focus of PD and technology integration was on teacher classroom practice and student learning. Finally, in all of the cases, we highlight the importance of teachers seeing students succeed through technology use.

Partnerships

Across the cases we saw varying levels of partnership between teacher and educational researchers or PD support providers. Anita worked closely with her research partner to identify solutions for a purpose larger than her own teaching context, Helen sustained a yearlong partnership with an instructional coach who support her use of online museum resources for history instruction, and Frank continued his working relationship with a now former teacher, which resulted in an innovative research study on the affordances of collaborative writing. This partnership approach presented a number of advantages, including the ability to quickly identify and understand teacher problems of practice, a detailed understanding of the teaching contexts, and how technology might best fit in each context to promote student learning. Additionally, this partnership approach included various forms of teacher feedback that allowed for more targeted support and more meaningful teacher engagement in their own learning. Additionally, this approach provided researchers and support providers with opportunities to quickly address teacher needs as they arose. Using short iterative, responsive feedback cycles that scaffolded teachers' development as users of instructional technologies were central to the partnership's as well as teachers' instructional development.

Focus on Teacher Problems of Practice and Student Learning

Consistent with prior research, by surfacing teachers' classroom problems of practice, teachers and researchers were able to work together to find productive ways to integrate technology into instruction in a manner that was both comfortable to the teacher and worked to improve student learning (Penuel, Fishman, Cheng, & Sabelli, 2011). Anita's concern about her students' frustration with "paging through the paragraphs" lead to the use of a keyboard shortcut for locating textual evidence in digital text. Helen, who initially expressed concerns about the lack of available technology in her classroom, was able to engage in student-centered instruction through a collaborative strategy and lesson designed in partnership with her coach. Interestingly, despite the focus on student learning outcomes, Frank ultimately chose to use the technology with a student population he deemed more appropriate. This decision required a deep understanding of his students' digital literacy skills at each grade level.

Often PD approaches tend to focus on tools or specific instructional approaches rather than attending to the problems of practice that teachers need addressed. This may result in limited to no implementation of the technology and little impact on student learning. In our cases, as the PD and supports focused on how to use the technology to address individual classroom challenges, we observed successful engagement and tool implementation through these partnerships. We also note the widely differing instructional challenges faced by these teachers and how a one-size-fits-all approach would have failed to address their individual needs.

Seeing Students Learn and Succeed.

Through these cases, we observed the importance of teachers seeing student succeed while learning with technology. We saw two teachers, Anita and Helen, persist or expand their use of technology based on seeing themselves and their students succeed through their respective digital tools. Through seeing her students' success and reflecting on the successful practices of a teacher-consultant, Anita explored and share with her colleagues a novel method of supporting students' oral fluency skills. Helen, with focused support from an instructional coach and by seeing her students' engagement, continued to use the online museum resources in more collaborative ways. On the other hand, despite seeing student success with his 8th graders, Frank was more optimistic that his older, more digitally literate college students would be better suited for using the online tool. Thus, although Frank continued to use the digital tool after the study, he used it with a different population of students. Nonetheless, all teachers ultimately used technology to promote student engagement and learning in new ways. They were also able to overcome challenges and tensions partly because they saw student success. We note that we only present three cases here. Further research should include examining a larger number of teachers to better understand the relationships between teachers seeing their students succeed and their persistence in using technological resources.

IMPLICATIONS FOR GLOBAL LEADERS BRIDGING DIGITAL DIVIDES

In this chapter, we detailed cases of leaders in research, teacher PD, and classroom instruction working together to bridge teacher digital divides. We saw that bringing together teachers and researchers who have a shared vision for student success helped springboard and sustain PD initiatives intended to address teachers' individual needs when using an instructional technology. We also saw that researchers and professional developers actively sought to identify teacher instructional needs in the classroom. This allowed the developers to find ways to integrate technology that aligned more closely with teachers' goals and needs. When technology was integrated successfully, teachers were able to see students succeed in using the technological tool, which likely increased teachers' confidence and investment in the tool. In turn, these teachers went on to experiment with and use these technologies in novel or expended ways.

In the case of Anita and her partner we saw their global leadership in the exploration of classroom practices that capitalize on the student learning affordances of 21st Century instructional technologies. In the case of Helen, we saw a partnership and shared leadership with a team of researchers and professional developers to design learning targeted for her needs. Frank and his colleague represent global leaders using forward thinking classroom practices and research that addresses digital divides through peer-to-peer online collaboration.

Though we presented three cases, there are some important implications for those who lead in the work of supporting teacher learning to bridge their digital divides. When planning or approaching teacher learning opportunities, it is critical to find ways to present and promote the use of technology as a way of addressing classroom instructional challenges, rather than presenting tools as solutions or as decontextualized resources. Along those lines, it is also important to position teachers so they quickly see their own success and the success of their students. By doing so, teachers may be more likely to remain engaged through the initial instructional experiences with the digital tool—a critical time period. Finally, it is important to remain flexible as teacher expertise develops and their instructional needs evolve. Thereby, research partners will be able to meet teachers where they are in their developmental trajectories.

REFERENCES

Ainley, J., Enger, L., & Searle, D. (2008). Students in a digital age: Implications of ICT for teaching and learning. In J. Voogt & G. Knezek (Eds.), *International handbook of information technology in primary and secondary education* (pp. 63–80). New York, NY: Springer

Cohen, E. G., & Lotan, R. A. (2014). *Designing groupwork: Strategies for the heterogeneous classroom third edition.* New York, NY: Teachers College Press.

Common Core State Standards Initiative. (2010). *Common core state standards for English language arts & Literacy in history/social studies, science, and technical subjects.* Washington, DC: National Governors Association Center for Best Practices and the Council of Chief State School Officers.

Cuban, L. (2013). *Inside the black box of classroom practice: Change without reform in American education.* Harvard Education Press.

Cuban, L., Kirkpatrick, H., & Peck, C. (2001). High access and low use of technologies in high school classrooms: Explaining an apparent paradox. *American Educational Research Journal, 38*(4), 813–834.

Darling-Hammond, L. (2017). Teacher education around the world: What can we learn from international Practice? *European Journal of Teacher Education,* 1–19.

Ertmer, P. A., Ottenbreit-Leftwich, A., & York, C. S. (2007). Exemplary technology-using teachers: Perceptions of factors influencing success. *Journal of Computing in Teacher Education, 23*(2), 55–61.

Epstein, D., Nisbet, E. C., & Gillespie, T. (2011). Who's responsible for the digital divide? Public perceptions and policy implications. *The Information Society, 27*(2), 92–104.

Fisher, D., & Frey, N. (2013). *Better learning through structured teaching: A framework for the gradual release of responsibility.* Alexandria, VA: ASCD.

Garet, M., & Porter, A. (2001). What makes professional development effective? Results from a national sample of teachers. *American Educational Research Journal, 38*(4), 915–945.

Incantalupo, L., Treagust, D. F., & Koul, R. (2014). Measuring student attitude and knowledge in technology-rich biology classrooms. *Journal of Science Education and Technology, 23*(1), 98–107.

Irvine, J. J. (2010). Foreword. In H. R. Milner's (Ed.), *Culture, curriculum, and identity in education* (pp. xi–xii). New York, NY: Palgrave Macmillan.

Isci, T. G., & Demir, S. B. (2015). The use of tablets distributed within the scope of FATIH Project for education in Turkey (is FATIH Project a fiasco or a technological revolution?). *Universal Journal of Educational Research, 3*(7), 442–450.

Koehler, M., & Mishra, P. (2009). What is technological pedagogical content knowledge (TPACK)? *Contemporary Issues in Technology and Teacher Education, 9*(1), 60–70.

Krishnan, J., Cusimano, A., Wang, D., & Yim, S. (2018). Writing together: Online synchronous collaboration in middle school. *Journal of Adolescent & Adult Literacy.* Retrieved from https://doi.org/10.1002/jaal.871

Krishnan, J., & Poleon, E. (2013). Digital Backchanneling: A strategy for maximizing engagement during a performance-based lesson on Shakespeare's Macbeth. *Teaching English with Technology, 13*(4), 38–48.

Mouza, C. (2009). Does research-based professional development make a difference? A longitudinal investigation of teacher learning in technology integration. *Teachers College Record, 111*(5), 1195–1241.

OECD (2016). *Innovating education and educating for innovation: The power of digital technologies and skills.* Paris, France: OECD Publishing.

Papert, S. (1990). Computer criticism vs. technocentrism. *Epistemology and Learning,* Massachusetts CA, (1).

Pearce, K. E., & Rice, R. E. (2013). Digital divides from access to activities: Comparing mobile and personal computer Internet users. *Journal of Communication, 63*(4), 721–744.

Penuel, W. R., Fishman, B. J., Cheng, B. H., & Sabelli, N. (2011). Organizing research and development at the intersection of learning, implementation, and design. *Educational Researcher, 40*(7), 331–337.

Peres, S. C., Tamborello, F. P., Fleetwood, M. D., Chung, P., & Paige-Smith, D. L. (2004, September). Keyboard shortcut usage: The roles of social factors and computer experience. In *Proceedings of the Human Factors and Ergonomics Society Annual Meeting* (Vol. 48, No. 5, pp. 803–807). Los Angeles, CA: Sage Publications.

Polly, D., & Orrill, C. H. (2016). Designing professional development to support teachers' TPACK in elementary school mathematics. In M. Herring, J. Koehler, & P. Mishra, (Eds.), *Handbook of technological pedagogical content knowledge* (pp. 259–270). New York, NY: Routledge.

Rosen, Y., & Beck-Hill, D. (2012). Intertwining digital content and a one-to-one laptop environment in teaching and learning: Lessons from the time to know program. *Journal of Research on Technology in Education, 44*(3), 225–241.

Tate, T. P., Warschauer, M., & Abedi, J. (2016). The effects of prior computer use on computer-based writing: The 2011 NAEP writing assessment. *Computers & Education, 101*, 115–131.

Thomas, M., Reinders, H., & Warschauer, M. (Eds.). (2012). *Contemporary computer-assisted language learning.* New York, NY: A&C Black.

van Deursen, A. J., & van Dijk, J. A. (2015). Toward a multifaceted model of internet access for understanding digital divides: An empirical investigation. *The Information Society, 31*(5), 37–391.

Van Dijk, J. A. (2017). Digital divide: Impact of access. *The International Encyclopedia of Media Effects*, 1–11.

Walker, S., Schloss, P., Fletcher, C. R., Vogel, C. A., & Walker, R. C. (2005a). Visual-syntactic text formatting: A new method to enhance online reading. *Reading Online, 8*(6), 1096–1232.

Walker, R. C., Schloss, P., Vogel, C. A., Gordon, A. S., Fletcher, C. R., & Walker, S. (2007). Visual-syntactic text formatting: Theoretical basis and empirical evidence for impact on human reading. In Proceedings of the IEEE International Professional Communication Conference. Seattle, WA.

Walker, R., Vogel, C., & Eagle, C. O. (2005b). *Live Ink®: Brain-based text formatting raises standardized reading test scores.* Philadelphia, PA: National Educational Computing Conference.

Wang, D. (2016). Exploring and supporting today's collaborative writing. In *Proceedings of the 2016 CHI Conference Extended Abstracts on Human Factors in Computing Systems* (pp. 255–259). Seoul, Korea: ACM.

Wang, D., Olson, J. S., Zhang, J., Nguyen, T., & Olson, G. M. (2015). DocuViz: Visualizing collaborative writing. In *Proceedings of the 33rd Annual ACM Conference on Human Factors in Computing Systems* (pp. 1865–1874). Seoul, Korea: ACM.

Warschauer, M., Cotten, S. R., & Ames, M. G. (2011). One laptop per child Birmingham: Case study of a radical experiment. *International Journal of Learning, 3*(2), 61–76.

Warschauer, M., Park, Y., & Walker, R. (2012). Transforming digital reading with visual-syntactic text formatting. *JALT CALL Journal, 7*(3), 255–270.

Warschauer, M., Zheng, B., & Park, Y. (2013). New ways of connecting reading and writing. *TESOL Quarterly, 47*(4), 825–830.

Yim, S., & Warschauer, M. (2017). Web-based collaborative writing in L2 contexts: Methodological insights from text mining. *Language Learning & Technology, 21*(1), 146–165.

Zinger, D., Naranjo, A., Naranjo A., Gilbertson, N., & Warschauer, M. (2017). A design-based research approach to improving professional development and teacher knowledge: The case of the Smithsonian Learning Lab. *Contemporary Issues in Technology and Teacher Education, 17*(3), 388–410.

Zinger, D., Tate, T., & Warschauer, M., (2017). Learning and teaching with technology: Technological pedagogy and classroom practice. In J. Candinin, & J. Husu (Eds.), *The SAGE handbook of research on teacher education* (pp. 577–593). London, UK: SAGE

CHAPTER 9

SOCIAL NETWORKING TECHNOLOGY AND THE SOCIAL JUSTICE IMPLICATIONS OF EQUITABLE OUTCOMES FOR FIRST-GENERATION COLLEGE STUDENTS

Yesenia Fernandez, Nancy Deng, and Meng Zhao

First-generation college student (FGCS) persistence and attainment has become key in ensuring social justice and equity on college campuses. However, a critical aspect of FGCS success, access to technology and social technology use as a tool, have not been central to the strategies implemented by post-secondary leaders to improve FGCS outcomes. Digital divide authors point out there is not only unequal access to technology but also differentiated use by socio-economic status when it comes to using it to access valuable information sources (DiMaggio et al, 2004; Hargittai, 2010). In addition, Martinez-Aleman, Rowan-Kenyon, and Savitz-Romer (2012), and Wohn et al (2013) discuss how social networking technologies can be tools to support FGCS persistence. Through this qualitative study informed by social capital and social network theory, we examine the type of social ties (family, friend, institutional) and institutional resources that social media use enables FGCS

Crossing the Bridge of the Digital Divide: A Walk with Global Leaders, pages 145–162.
Copyright © 2019 by Information Age Publishing

to strengthen or build. We discuss the promising use of social networking technology to expand students' access to information networks and social capital which is critical to persistence. Finally, we underscore how the equitable access to technology has social justice implications when it comes to the post-secondary outcomes of first-generation college students.

INTRODUCTION

For decades, researchers have studied the experiences of first-generation college students (FGCS) in hopes of better understanding their trajectory and deciphering best practices to mitigate the unequal outcomes students experience in higher education. Yet, access to technology and social technology use as a tool are nearly absent from the extant literature related to FGCS post-secondary outcomes. Researchers such as Martinez-Aleman, Rowan-Kenyon, and Savitz-Romer (2012), Ellison, Steinfield, and Lampe (2007), and Wohn, Ellison, Khan, Fewins-Bliss, and Gray (2013) have highlighted the additive role of online social network technologies on college campuses and upon persistence of first-generation college students. In addition, digital divide authors point out there is differentiated use of technology by socio-economic status, particularly when it comes to what they term "capital enhancing" activities or activities that will enable access to valuable information sources (DiMaggio, Hargittai, Celeste, & Shafer, 2004; Hargittai, 2010). As such, the intersection of technology use and FGCS persistence is critical to our examination of how to support FGCS throughout the post-secondary pipeline and improve social justice and equity on college campuses. For the purposes of this discussion, we operationalize FGCS as students whose parents' highest level of education is a high school diploma or less (Chen, 2005; Nunez & Cuccaro-Alamin, 1998). First-generation college students are the first in their families to navigate post-secondary institutions in the US and consequently, must decode institutional practices and expectations largely on their own.

Engle and Tinto (2008) point out, up to 43% of low-income FGCS do not persist after six years and in an earlier study, Chen and Carroll (2005) state that while 68% of middle-class students graduated within 8 years, only 24% of low-income students graduated within that timeframe. Most recently, a 2015 Pell Institute report entitled "Indicators of Higher Education Equity in the United States," reported that only 9% of students from low-income households earned a bachelor's degree by age 24 compared to 77% of students from high income households. This income gap in bachelor's degree completion is greater than it was 43 years ago according to the same report. An aim of social justice leadership is equitable schooling and education (Jean-Marie, Normore, & Brooks, 2009; see also Normore, 2008) and as low income and FGCS continue to lag behind high income and continuing-generation college students in degree completion, it is a matter of social justice for educators to develop strategies to ensure equity. Furthermore, an important aspect of being a global leader in higher education is the idea of ethnorelativism or adapting to our global context and understanding the cultures of the students we serve and thus adjusting our strategies to improve their academic outcomes (see Marshall, 2015's discussion of global leadership in higher education). Knowing this, we set out to examine how first-generation college students use technology, social media in particular, as a tool as they navigate these complex institutions. Our qualitative study of 109 students at an urban four-year public university provides insight into college students' use of technology and social media. In our discussion, we consider how these may be instrumental to improve FGCS persistence and we also underscore the importance of broadening the digital divide conversation to include its intersection with FGCS persistence as this has social justice implications when it comes to post-secondary outcomes.

FIRST GENERATION COLLEGE STUDENTS AND INEQUITABLE OUTCOMES

According to Redford, Hoyer, and Ralph's (2017) US Department of Education's National Center for Education Statistics (NCES) report, of high school sophomores in 2002 who later enrolled in a post-secondary institution, roughly 24% were first-generation college students. The extensive report compares the characteristics of FGCS and continuing-generation students (students whose parents have at least a bachelor's degree) and notes that college attainment among FGCS compared to continuing-generation college students "...is unequally distributed." Lauff and Ingels (2013, as cited in Redford, Hoyer, & Ralph, 2017) found that,

> among 2002 high school sophomores, 46 percent of students who had a parent with a bachelor's degree and 59 percent who had a parent with a master's degree or higher had obtained a bachelor's degree or higher by 2012, compared to 17 percent of students who had parents with no postsecondary education experience (or "first-generation" college students (p. 1).

That is approximately a 30-percentage point difference in degree completion between FGCS and continuing-generation college students. Engle's earlier work (2007) also pointed out "...first-generation students [are] much more likely to leave (29 versus 13 percent) and much less likely to earn a degree (47 versus 78 percent) than students whose parents had a college degree" (p. 26). The NCES report also explained that more of the FGCS came from low-income households compared to continuing-generation students, "...households making $20,000 or less (27 vs. 6 percent) and $20,001 to $50,000 (50 vs. 23 percent)" (p. 4). Moreover, "...52 percent of such students first enrolled in 2-year institutions" (p. 9). This is important to note since 2-year institutions are the least selective of the higher education options and students often do not graduate or transfer even after spending 3–4 years there according to the report. Other authors have also mentioned the fact that most FGCS enroll at 2-year institutions although they transfer or complete a degree at low rates (Adelman, 1999; Ishitani, 2003, 2006; Jenkins & Fink, 2016; Striplin, 1999). Indeed, it has been well documented that FGCS are less likely to persist in higher education than continuing-generation college students and that our institutions produce unequal outcomes (see also, Adelman, 1999; Chen, 2005; Lohfink & Paulsen, 2005; Pascarella, Pierson, Wolniak, & Terenzini , 2004; Pike & Kuh, 2005).

Much of the conversation regarding the reasons for disproportionate outcomes has centered on precollege factors. Researchers have analyzed the impact that poverty and parent educational level have on FGCS persistence citing both as factors that negatively affect persistence and which put FGCS at a disadvantage (Choy, 2001; Engle & Tinto, 2008; Nunez & Cuccaro-Alamin, 1998; Thayer, 2000; Walpole, 2003). In addition, high school course enrollment or academic preparation has also been identified as a critical aspect of FGCS ability to successfully persist in higher education. The more rigorous the coursework and higher grade point average the more prepared students are and the greater likelihood they have to persist, according to these researchers (Adelman, 1999; Chen & Carrol 2005; Horn & Kojaku, 2001; Riehl, 1994; Striplin, 1999; Terenzini, Rendon, Upcraft, Millar, Allison, Gregg, Jalomo, 1994; Warburton, Bugarin & Nunez, 2001). The pre-college factors outlined (poverty, parent education level, and taking rigorous coursework in high school) are ultimately factors which "blame the victim", as Tinto (2006) points out, and have largely dominated the focus of postsecondary institutions searching for strategies to improve persistence rates. As a result of this research, post-secondary institutions have operationalized findings by creating programs which support students academically and financially. Some of these programs to support FGCS have included remedial coursework, summer bridge, first-year program, as well as access to technology and financial support (Engle & Tinto, 2008; Ishitani, 2006; Lohfink & Paulsen, 2005; Stephens et al., 2015; Thayer, 2000; Tinto, 2006).

Nevertheless, as Ishitani (2006) states, "The greatest benefits for explaining college success of first-generation students result from thorough examination of both precollege attributes of students and the quality of their interactions with institutions of higher education" (p. 865). How students interact with institutions and the college community, such as how engaged, connected, and academically and socially integrated they are, have been cited as critical to FGCS persistence in higher education (Gloria & Castellanos, 2012; Pascarella, 1980, Pascarella et al., 2004; Rios-Aguilar & Deil-Amen, 2012; Stephens, Fryberg, Markus, Johnson, & Covarrubias, 2012; Terenzini & Pascarella, 1980; Tinto, 1998). Pascarella et al. (2004) explains the research related to student persistence which focuses on student integration and social and academic engagement within the college community. This work includes Tinto's (1975, 1987) student integration framework, a model of student retention. Pascarella et al. (2004) states,

> Since first-generation students are likely to enter college with a lower stock of cultural/social capital than their peers, one might anticipate that their levels of academic, and perhaps even social, engagement during college will function in ways that may help them make up for this deficit. That is, levels of academic and social engagement will act in a compensatory manner... (p. 252).

This research is based on the idea that integration and engagement in college translates into social capital and social networks for students to utilize to successfully navigate post-secondary institutions (see Pascarella & Chapman, 1983; Pascarella & Terenzini, 1983). Rios-Aguilar and Deil-Amen (2012) study how social networks impact the trajectory of Latina students in higher education and define social capital as "…contacts and memberships in networks which can be used for personal gain…and it is typically thought of a resource that individuals exchange and accumulate" (p. 180). These resources can be found in social networks or groups of people who share a tie to one another (see also Coleman, 1988; Granovetter, 1973; Lin, 1999). In addition to Coleman, Lin, and Granovetter, Stanton-Salazar (1997) also discusses this theoretical framework and is cited in Martinez-Aleman et al. (2012) who assert that,

> Social capital theory, a theoretical framework that emphasizes the benefits of social networks, has been used by several researchers to illustrate the importance of relational support in preparing underserved students for college access and success (p. 13).

Essentially, researchers assert that the interactions FGCS have with their peers, faculty and other institutional agents, as well as their participation in out-of-classroom activities determine how integrated and engaged they are to the college community and this academic and social integration improves their likelihood to persist. Furthermore, according to Terenzini and Pascarella (1998) technology can play a critical role in this process as it can, "provide vehicles for active student involvement" (p. 159). Martinez-Aleman et al. (2012) study the intersection of social networking technology and FGCS success to gain insight into how it can be used as a tool to improve students' engagement and persistence. Similarly, DeAndrea, Ellison, LaRose, Steinfield, and Fiore (2012) take a look at how social media can be used to help underrepresented students adjust to college. How technology use (not simply access to technology) can mitigate inequities experienced by marginalized communities is not only the focus of emerging literature by digital divide researchers but should also be a focus for equity minded global leaders working to improve outcomes for first generation college students.

FIRST GENERATION COLLEGE STUDENTS AND THE DIGITAL DIVIDE

Goode (2010) explains there has to be a redefining and reconceptualizing of the digital divide literature as it is more about how technology is used, the sociocultural context, and how it may engender social inclusion and social networks and the resources these bring. Other digital divide authors have echoed this as our global context engenders a more complex relationship with technology than simply having access to it (see Dolan, 2016; Eubanks, 2012; Gorski, 2009; Payton, 2003; Valadez & Duran, 2007; Warschauer, 2002, 2004). For example, Goode's

(2010) study documents "technobiographies" or narratives of five low-income and three high-income high school students' relationship with technology in K-12 and provides insight into the existing digital inequities low-income students experience. Interviews with teachers revealed that high income students were more likely to analyze data, carry out research, produce multimedia projects and create demonstrations with computers than low-income students. What Goode ultimately explains is that these students' technobiographies "demonstrate how holding a particular technology identity creates both academic opportunities and obstacles for students" (p. 509). This becomes particularly poignant as Goode compares students' relationship with technology or rather how they use it in K-12, with how they use it as they enter post-secondary institutions. Once students in this study entered post-secondary institutions, they used technology in the same way as they did in high school and there was nothing that the institution did to change this. This is an important finding as most FGCS are low-income students and this research suggests it is up the institution to make an impact on the technology knowledge they bring with them to higher education. Indeed, as Goode concludes, "Technology knowledge serves as a gatekeeper to college success" (p. 505).

Sax, Ceja, and Teranishi (2001) also found disparities in pre-college use of information technologies by race, class and gender. They highlight how lived experiences play a role in creating and exacerbating the digital divide beyond access. Sax, Ceja, & Teranishi contend that it is not so much about unequal access to technology but rather differentiated use by socio-economic status which impacts students' ability to gain academic success. Dolan (2016) similarly advocates for a conceptual shift of the digital divide from "material access" of technology to its actual use. Furthermore, Wei and Hindman (2011) write about the relationship between the digital divide and the knowledge gap and state, "the digital divide… can be better defined as inequalities in the meaningful use of information and communication technologies" (p. 217). These authors study the impact of internet use on people's knowledge of social and political discussions and outline how depending on sociocultural factors, people use information and communication technologies to gain knowledge and impact the knowledge gap. Similarly, Hargittai (2010) argues that internet usage patterns differ by sociocultural context as the web skills of lower socio economic communities were not as sophisticated as those of more affluent ones. This digital divide, as defined by the inequities in the utilization of technology as a tool, leads to further stratification of low income communities and vulnerable populations. All of the aforementioned authors, are pointing out how the digital divide is not about access to technology but rather it's how the use of technology "…reflects and reinforces society's social and economic inequities" (Goode, 2010, p. 500; see also Light, 2001; Warschauer, 2004). Because of its potential to reinforce existing inequities, we must examine the digital divide through the lens of how the technology is used or not used by underrepresented student groups which is what we have attempted to do with the study discussed in this chapter.

OUR STUDY: SOCIAL NETWORKING TECHNOLOGY AND FIRST GENERATION COLLEGE STUDENTS

Methods

This qualitative study consisted of an online survey administered to students from two colleges, the College of Business Administration & Public Policy (CBAPP) and the College of Education (COE) at an urban public university. Our aim was to determine the social and academic support students receive from peers, family and institutional agents (students' social network), how students use social media platforms, as well as what students' opinions are of their usefulness. We compared the survey responses of FGCS with continuing-generation college students' responses to gain insight into the differences in their ties or networks, their use of technology, and in their opinions of its academic and social utility. Over the course of two months and after one pilot which involved 19 students, the revised survey was made available to students from the aforementioned two colleges. Researchers consequently collected data from 107 students from a total of 5 classes. There were 80 responses (74.8% of 107) from CBAPP students, and a total of 96 responses (out of the 107 participants) were complete and are used in the data analysis included in this narrative.

The first part of the survey was comprised of multiple choice questions in order to gather demographic data. To determine which students were FGCS, there was a question which asked participants to identify themselves as "first-generation students" based on listed criteria. There were 73 students who identified as FGCS (76%) and 23 of the participants did not identify as FGCS (they were continuing-generation college students) and will be referred to as Non-FGCS in this narrative. Then, the rest of the survey was comprised of some open-ended questions and some multiple questions. The first open-ended question asked participants about the people who provide them with academic support (their interactions with peers, family, institutional agents). Students were asked:

> Who do you go to for help on your academic work? Please name the person/ relationship (this could be anyone-family, friends, professor, advisor, counselor) and briefly explain what kind of help they give you.

Multiple choice questions asked students about their social media use and included questions regarding:

- the social media platforms used
- time spent on social media
- frequency of using social media

Finally, students responded to open-ended questions which asked them to share their experiences and opinions related to the utility of their social media use:

1.   Do you think social media (Snapchat, Twitter, Instagram, Facebook etc.) helps your social interactions (e.g., with family, friends, classmates, instructors, school staff)? Why or why not?
2.   Do you think social media (Snapchat, Twitter, Instagram, Facebook etc.) helps you academically? Why or why not?

For the first open-ended question "Who do you go to for help on your academic work?" students could list as many people as they wanted and thus researchers coded responses as follows:

1.   Family
2.   Peer
3.   Professor
4.   Other Institutional Agent
5.   Outside Institution
6.   No one. Researchers use descriptive statistics (frequencies and percentages) to report findings in this narrative.

FINDINGS

The majority of the students who took the survey identified as Hispanic/Latino (61%), were seniors (52%), and were employed (84%). The percentages of the responses for ethnicity, majors, and student academic standing are listed in Table 9.1 below. Most of the students reported majors that were in the business field while 21% reported liberal studies or graduate education as their major.

In response to the question "Who do you go to for help on your academic work?" students were able to identify as many people as they wanted and researchers tallied responses using the coding previously listed. Almost half of the FGCS (48%) identified peers as one of the persons from whom they receive academic support compared to 35% of Non-FGCS. Fifteen FGCS identified peers as the ONLY person who helps them with academic work (that's 21 percent). Only 4 FGCS identified family alone while 8 total students (11%) identified a family member along with someone else as the person who helps them with academics. This is compared to 49% of Non-FGCS (11 students out of the 23 total respondents) who reported they receive help with their academic work from family. A substantially higher percentage of Non-FGCS turn to family for help on academic work than FGCS. Eleven percent of FGCS identified professors as the ONLY person they go to for help with academics and 38% named a professor along with other people. On the other hand, only 11% of Non-FGCS identified a professor as the person they go to for help. Based on these findings, for FGCS, professors are critical to assist their academic trajectory. Actually, institutional agents seem to be very important to FGCS based on these responses. In fact, 32% of FGCS identified their advisor or an on-campus tutoring center as one of the people to whom they go for help on academics (30% for Non-FGCS). But what is most telling, 14

TABLE 9.1. Sample Characteristics (n=96)

Ethnicity	Hispanic or Latino	61%
	Asian or Pacific Islander	18%
	Black or African American	10%
	White/Caucasian	8%
	Other	2%
Academic Standing	Seniors	52%
	Graduate	24%
	Juniors	24%
Employment Status	Employed Part-time	43%
	Employed Full-time	41%
	Not Employed	16%
Majors	General business administration	26%
	Global Logistics and Supply Chain	14%
	Graduate Program-Education	13%
	Liberal Studies	8%
	Criminal Justice	7%
	Accounting	6%
	HR Management	6%
	Entrepreneurship	3%
	Marketing	3%
	Sports Entertainment & Hospitality	3%
	Sociology	3%
	Others	7%

students of the 32% who identified their advisor or tutor said that's the ONLY person they go to for help. That means 19% of all FGCS said they ONLY rely on their advisor for academic help. Thus, peers, professors and other institutional agents (advisors) came up as the only person FGCS go to for help for 51% of the respondents and for 95% of FGCS, they were one of the multiple persons on campus they go to for help. Two out of the FGCS identified YouTube and "the internet" as the places they go to for help, and another 3 students first said "no one" or only themselves, but then identified someone like a professor, tutor or peer. Ultimately, FGCS rely on institutional agents for academic support the most.

In an effort to determine how social media is used by FGCS compared to Non-FGCS, the survey then asked participants to list *one major* social media account they have used. Instagram (38%) and Facebook (36%) were the social media platforms most frequently identified by survey respondents while LinkedIn, Twitter, and Pinterest were the social media platforms least frequently identified. Insta-

TABLE 9.2. Major Social Media Platform Used

Name of Platform	Percentage
Instagram	38%
Facebook	36%
Snapchat	10%
LinkedIn	8%
Twitter	5%
Pinterest	3%

gram and Facebook were identified more than three times as much as the other social media platforms mentioned. Table 9.2 reveals the frequency with which students listed each major social media platform.

More than 80% of students who responded to the survey noted they spend an hour or more on social media on a typical day. Non-FGCS reported to spend more hours on social media than FGCS (43% reported they spend 3–6 hours); however, only FGCS (4% of them) reported they spent 7 hours or more on social media and a total of 35% reported they spend 3–7 hours on social media. The question students were asked specifically was, "In a typical day, how much time do you use the social media?" and Table 9.3 provides a comparison of students' responses. It is clear from these responses that college students spend a substantial amount of their day on social media.

Students were also asked how often, or frequently, they log onto their social media platform and were given options ranging from not at all, to every 7 or more hours. The majority (over 60%) of both the Non-FGCS and FGCS survey respondents reported they log on or use their social media platform every 2–3 hours or less. That means that not only do they spend an inordinate number of hours on social media platforms once logged on, but they also only spend, at most, 3 hours off the platforms. Indeed, 32% of FGCS responded that they use social media every

TABLE 9.3. Social Media Use

	Non-FGCS (n=23)	FGCS (n=73)	All (n=96)
Q: In a typical day, how much time do you use the social media?			
None	9%	15%	4%
Less than 1 hour	9%	3%	14%
1–2 hours	39%	47%	45%
3–4 hours	30%	19%	22%
5–6 hours	13%	12%	13%
7 hours or more	0%	4%	3%

TABLE 9.4.   Social Media Use in the Sample (n=96)

	Non-FGCS (n=23)	FGCS (n=73)	All (n=96)
Q: In a typical day, how often do you use the social media?			
Not at all	9%	5%	6%
Every 30 mins	22%	10%	13%
Every 1 hour	13%	22%	20%
Every 2–3 hours	26%	33%	31%
Every 4–6 hours	17%	18%	18%
Every 7 hours +	13%	12%	13%

1 hour or every 30 minutes. Table 9.4 lists the frequencies as reported by survey participants. It is undeniable, based on these responses, that our college students spend a large part of their day logged onto social media. For some respondents, almost all of their waking hours are spent on social media. Thus, we can argue these students have access to technology. The question remains, then, what do they do with it? And is what they do, useful to their ability to navigate higher education.

The open-ended questions in the survey then provided researchers insight into students' perspectives regarding the usefulness of social media. As listed previously, we asked students if they thought social media helped them academically and socially (to be integrated and engaged which the previously discussed literature suggests can improve student outcomes). Student responses showed the use of social media platforms was useful for students to build upon their social network and gain information that was important to their college coursework. Students were able to use technology, a social media platform in this case, to make or maintain connections, be involved on campus and thus be integrated. Some students also recognized social media as a tool, for example, to organize meetings and events, study groups, and keep up in classes. Table 9.5 shows some of the students' responses related to the use of social media that impacted their integration and engagement within the network of the campus. Furthermore, students also recognized the use of social media platforms as a tool for academic support which could help them gain information that could assist them in their classes. Some respondents commented they could be used to support them in efforts such as to be informed, write papers, and collaborate with other students. Table 9.6 has examples of those responses.

Interestingly, participants' responses to the question "Do you think social media helps you academically?" varied by age group and FGCS status. More students between the ages of 1–23 perceived social media as a distraction and not as a tool which could benefit them academically. While 57% of students ages 1–23 regarded social media use as a distraction, only 44% of students aged 24 or higher felt social media was a distraction and not useful academically. In addition, 77%

TABLE 9.5. Social Media and Social Networks

Student Demographics	Response
Q: Do you think social media (Snapchat, Twitter, Instagram, Facebook etc.) helps your social interactions?	
20 years old FGCS	Of course. I follow various school pages, I find out about jobs, events, and it helps us communicate with people we cannot meet face to face.
22 years old Non-FGCS	It is easy to socially interact through social media and it is easier to contact people sometimes. For example, you can DM a person through Instagram and hold a private conversation and you can do the same with Facebook, etc
22 years old FGCS	Yea, that's what all the kids like to do. Even when I met someone in my class and got to know each other better, she went straight in and asked "what's your snapchat?" So it does help and a great resource to stay in touch with others.
23 years old FGCS	I think so because there are some friends that we don't see as often but keep in touch through social media
25 years old Non-FGCS	Social media helps my social interactions with new friends. Everyone communicates online now and hardly picks up the phone anymore
26 years old Non-FGCS	Social media allowed for me as a club president to keep in touch with the rest of my executive board.
26 years old FGCS	Yes, because you can create plans to meet people through social media.
28 years old Non-FGCS	Sometimes. It can be helpful once you have met someone to keep in contact with them for school or work purposes.
29 years old FGCS	Yes, sometimes instead of giving my number to study group I give them my user name instead.
30 years old, Non-FGCS	Helps me network

TABLE 9.6. Social Media and Academic support

Student Demographics	Response
Q: Do you think social media (Snapchat, Twitter, Instagram, Facebook etc.) helps you academically?	
19 years old FGCS	It does help academically because for example if you miss a day, it can serve as communication between classmates to get the work you missed.
20 years old FGCS	Not really...unless the class or group of students get together to communicate through one of those platforms. Otherwise, it's just another distraction.
21 years old Non-FGCS	Yes if it includes educational information, which sometimes does
22 years old Non-FGCS	Yes. Helps with looking up hard to find answers
25 years old FGCS	I can see what struggles peers are having, browse teaching videos, get planning ideas, and look at pictures of projects teachers have tried.
29 years old FGCS	I find a lot of good articles to read that are related to my career, and also ideas for my classroom.
41 years old FGCS	Yes. I follow organizations that post educational material. I also look at ideas from other educators to give me ideas to support students.

of respondents between the ages of 1–23 were FGCS and of those students, 57% said social media was a distraction and could not be used as a tool to help them academically. Similarly, from the entire sample of FGCS (all ages) who took the survey, 57% responded that social media use was a distraction and not useful to them academically. It appears from our findings that more of the older students had figured out how social media use could be a tool and not a waste of time but that FGCS regardless of age group, had not.

DISCUSSION

Findings indicate that FGCS who took this survey primarily rely on institutional agents for academic support. One FGCS explained "I have no one in the family that can help, but I try to look for professor's advice" and another FGCS said they "got over feeling embarrassed" and went to their professor for assistance on their academic work. Another FGCS said, "Peers…have been crucial in my success. We talk about assignments, remind each other when they're due, show each other our work, suggestions, etc." Rios-Aguilar and Deil-Amen (2012), Stanton-Salazar (1997, 2011) as well as Pascarella (1980) have highlighted how critical these ties are for students to have within their social network. Moreover, although FGCS rely primarily on institutional agents and not family for assistance on academic work, one of the FGCS commented that their family members "help keep life together" and those who reported they get help from their family explained that their parent was also in school. This is consistent with extant literature on the benefits of familial support for FGCS (Dennis, Phinney & Chuateco, 2005; Gloria & Castellanos, 2012; Perna & Titus, 2005; Wohn et al. 2013). Upon analysis of responses for both questions regarding social media use and social network development and academic support we found some students were able to use social media as a tool to help them academically and to expand or strengthen their social network on campus as well as gain academic support. Survey responses indicate that social media has provided important spaces which are instrumental in improving FGCSs' social capital critical to their success (DeAndrea et al. 2012; Ellison, Steinfield, & Lampe, 2007; Martinez-Aleman et al. 2012; Wohn et al, 2013). The dual effects of social media use on students' academic work was also an important finding. While the use of social media mostly have positive effect on social capital development and academic support, social media caused distraction to academic work and that overshadowed its usefulness for most of the survey participants. As Wohn et al. (2013) explains, social media can broaden a students' network of people who can provide support in the form of information but often this is not the case for FGCS as they haven't realized how to use it as a tool for that. This was also a finding in (Hargittai, 2010; Rios-Aguilar & Deil-Amen, 2012; Wohn et al., 2013). The fact that FGCS regardless of age group had not figured out how social media use could be a tool and not a waste of time is an important finding as it highlights the importance of intentionally teaching students this skill to support their trajectory in higher education. Some of our findings are similar to Goode's

(2010) study in which he concluded that those who have had interactions with technology, in this case social media, that have equip them, feel in control and empowered by it while those who have had less sophisticated interactions with it felt it was a distraction. Specifically, what Goode (2010) underscores in his discussion of the difference in technology knowledge by socio-economic status about students' technology knowledge and Hargittai (2010)'s comments on the lack of "capitalenhancing" behaviors using technology by marginalized groups cannot be ignored by post-secondary institutions. Virtually no student reported being required to use social media as a tool in their coursework by professors or by academic advisors or any other institutional agent. The university provides opportunities for those who already can use technology (in this case social media) as a tool but not intentional support for those who have not learned how to use it as a tool. What survey responses ultimately suggest is that how the university uses this social networking technology can reinforce existing inequities or bridge the knowledge gap that exists.

DIGITAL EQUITY AND IMPLICATIONS
FOR GLOBAL LEADERSHIP

If institutions are to close the tremendous achievement gap between Non-FGCS and FGCS, they must not simply focus strategies to mitigate precollege factors. Post-secondary institutions have an opportunity to capitalize on students' inordinate use of social networking technology to support FGCS through their trajectory. The digital divide is no longer about students' access to technology but rather on how that technology is used, technology knowledge and how social networking technology can be used as a tool to close the knowledge gap. We must teach our FGCS how to use technology in complex ways by being intentional in teaching them of its utility. This strategy can create opportunities for campuses to develop communities where FGCS are more integrated, connected and engaged and ultimately, expand students' access to social capital and social networks that are critical to their ability to navigate the institution and persist in higher education.

Leaders can address the global context of their institutions which frequently serve multicultural and FGCSs by developing systematic approaches for the use of social media platforms on their campus. The impact of sociocultural challenges upon the persistence of FGCSs in post-secondary institutions is one force global leaders must not ignore as it is a matter of social justice for us to develop systems that will support students through the higher education pipeline. As some of the responses by participants in this study reveal, FGCS often feel isolated and alone as they are the first in their families to navigate a post-secondary institution and they must rely upon strangers-institutional agents for academic support. Indeed, global leaders must recognize the interdependence between institutional agents, students and the critical information necessary to FGCS success (see Osland et al, 2006). It is critical if we are to ensure equity and justice on our campuses that systems thinking is applied to the use of technology tools and social networking platforms in particular that is inevitable in a global society. Some of the practices which can be systematized on post-secondary campuses can even be those which students in this study suggest such as being intentional in the use of social networking tools for study groups, communication with professors, information pertinent to courses, and providing students with information related to on campus events.

It is the challenge of global leadership for educational leaders whose goal it is to ensure equity and social justice on our campuses to approach student achievement not from a deficit perspective which blames students for sociocultural forces but rather from a global mindset which recognizes the complex environment and devices systems to address it.

REFERENCES

Adelman, C. (1999). *Answers in the toolbox: Academic intensity, attendance patterns, and bachelor's degree attainment.* Washington, DC: U.S. Department of Education, Office of Educational Research and Improvement.

Coleman, J. S. (1988). Social capital in the creation of human capital. *American Journal of Sociology, 94,* S95–S120.

Chen, X, (2005). *First-generation students in postsecondary education: A look at their college transcripts.* Washington, DC: National Center for Education Statistics.

Chen, X., & Carroll, C. D. (2005). *First-generation students in postsecondary education: A look at their college transcripts. postsecondary education descriptive analysis report.* NCES 2005–171. National Center for Education Statistics.

Choy, S. (2001). *Students whose parents did not go to college: postsecondary access, persistence, and attainment. findings from the condition of education,* 2001. INSTITUTION National Center for Education Statistics (ED), Washington, DC.; MPR Associates, Berkeley, CA. REPORT NO NCES-2001-126

DeAndrea, D. C., Ellison, N. B., LaRose, R., Steinfield, C., & Fiore, A. (2012). Serious social media: On the use of social media for improving students' adjustment to college. *The Internet and higher education, 15*(1), 15–23.

Dennis, J. M., Phinney, J. S., & Chuateco, L. I. (2005). The role of motivation, parental support, and peer support in the academic success of ethnic minority first-generation college students. *Journal of College Student Development, 46*(3), 223–236.

DiMaggio, P., Hargittai, E., Celeste, C., & Shafer, S. (2004). Digital inequality: From unequal access to differentiated use. In K. Neckerman (Ed.), *Social inequality* (pp. 355–400). New York, NY: Russell Sage Foundation.

Dolan, J. E. (2016). Splicing the divide: A review of research on the evolving digital divide among K–12 students. *Journal of Research on Technology in Education, 48*(1), 16–37.

Engle, J. (2007). Postsecondary access and success for first-generation college students. *American Academic, 3*(1), 25–48.

Engle, J., & Tinto, V. (2008). *Moving beyond access: College success for low-income, first- generation students. The Pell Institute for the study of opportunity in higher education.* Washington, DC: Author.

Ellison, N. B., Steinfield, C., & Lampe, C. (2007). The benefits of Facebook "friends:" Social capital and college students' use of online social network sites. *Journal of Computer- Mediated Communication, 12*(4), 1143–1168.

Eubanks, V. (2012). *Digital dead end: Fighting for social justice in the information age.* MIT Press.

Granovetter, M. (1973). The strength of weak ties. *American Journal of Sociology, 78,* 1360– 1380.

Gloria, A. M., & Castellanos, J. (2012). Desafíos y bendiciones: A multiperspective examination of the educational experiences and coping responses of first-generation college Latina students. *Journal of Hispanic Higher Education, 11*(1), 82–99.

Goode, J. (2010). The digital identity divide: how technology knowledge impacts college students. *New Media & Society, 12*(3), 497–513.

Gorski, P. C. (2009). Insisting on digital equity: Reframing the dominant discourse on multicultural education and technology. *Urban Education, 44*(3), 348–364.

Hargittai, E. (2010). Digital Na (t) ives? Variation in internet skills and uses among members of the "net generation". *Sociological inquiry, 80*(1), 92–113.

Horn, L., & Kojaku, L. K. (2001). High school academic curriculum and the persistence path through college: Persistence and transfer behavior of undergraduates 3 years after entering 4-year institutions. *Education Statistics Quarterly, 3*(3), 65–72.

Ishitani, T. T. (2003). A longitudinal approach to assessing attrition behavior among first-generation students: Time-varying effects of pre-college characteristics. *Research in Higher Education, 44*(4), 433–449.

Ishitani, T. T. (2006). Studying attrition and degree completion behavior among first-generation college students in the United States. *The Journal of Higher Education, 77*(5), 861–885.

Jean-Marie, G., Normore, A. H., & Brooks, J. S. (2009). Leadership for social justice: Preparing 21st century school leaders for a new social order. *Journal of Research on Leadership Education, 4*(1), 1–31.

Jenkins, D., & Fink, J. (2016). Tracking Transfer: New measures of institutional and state effectiveness in helping community college students attain bachelor's degrees. *Community College Research Center, Teachers College, Columbia University.*

Light, J. S. (2001). Rethinking the digital divide. *Harvard Educational Review, 71*(4), 709–33.

Lin, N. (1999). Social networks and status attainment. *Annual Review of Sociology, 25,* 467–487.

Lohfink, M. M., & Paulsen, M. B. (2005). Comparing the determinants of persistence for first- generation and continuing-generation students. *Journal of College Student Development, 46*(4), 40–428.

Marshall, V. L. (2015). *An exploration of global leadership practices implemented by successful higher education faculty members.* Dissertation. Lamar University-Beaumont.

Martínez-Alemán, A. M., Rowan-Kenyon, H. T., & Savitz-Romer, M. (2012). Social Networking and first-generation college student success: A conceptual framework for "critical" engagement and persistence efforts. Unpublished manuscript, Harvard Graduate School of Education, Cambridge, MA.

Normore, A. H. (Ed.). (2008). *Leadership for social justice: Promoting equity and excellence through inquiry and reflective practice.* Charlotte, NC: IAP.

Nunez, A. M., & Cuccaro-Alamin, S. (1998). *First-generation students: Undergraduates whose parent never enrolled in postsecondary education* (NCES 1999–082). Washington, DC: U.S. Government Printing Office.

Osland, J. S., Bird, A., Osland, A., & Mendenhall, M. (2006). *Developing global leadership capabilities and global mindset: a review*. In G. B. I. Stahl (Ed.), *Handbook of research in international human resource management* (pp. 197–222). Northampton, MA: Edward Elgar Publishing.

Pascarella, E. T. (1980). Student-faculty informal contact and college outcomes. *Review of Educational Research, 50*(4), 545–595.

Pascarella, E. T., & Chapman, D. (1983). A multi-institutional path analytical validation of Tinto's model of college withdrawal. *American Educational Research Journal, 20*(1), 87–102.

Pascarella, E. T., & Terenzini, P. (1983). Path analytic validation of Tinto's model. *Journal of Educational Psychology, 75*(2), 215–226.

Pascarella, E. T., Pierson, C. T., Wolniak, G. C., & Terenzini, P. T. (2004). First-generation college students: Additional evidence on college experiences and outcomes. *The Journal of Higher Education, 75*(3), 24–284.

Payton, F. C. (2003). Rethinking the digital divide. *Communications of the ACM, 46*(6), 89–91.

Perna, L., & Titus, M. (2005). The relationship between parental involvement as social capital and college enrollment: An examination of racial/ethnic group differences. *Journal of Higher Education, 76*, 485–518.

Pike, G. R., & Kuh, G. D. (2005). First-and second-generation college students: A comparison of their engagement and intellectual development. *The Journal of Higher Education, 76*(3), 276–300.

Redford, J., & Hoyer, K. M. (2017). *First-generation and continuing-generation college students: A comparison of high school and postsecondary experiences*. (ED-IES-12-D-0002.) U.S. Department of Education, National Center for Education Statistics. Washington, DC: U.S. Government Printing Office.

Riehl, R. J. (1994). The academic preparation, aspirations, and first-year performance of first- generation students. *College and University, 70*(1), 14–19.

Rios-Aguilar, C., & Deil-Amen, R. (2012). Beyond getting in and fitting in: An examination of social networks and professionally relevant social capital among Latina/o university students. *Journal of Hispanic Higher Education, 11*(2), 17–196.

Sax, L. J., Ceja, M., & Teranishi, R. T. (2001). Technological preparedness among entering freshmen: The role of race, class, and gender. *Journal of Educational Computing Research, 24*(4), 363–383.

Stanton-Salazar, R.D. (1997). A social capital framework for understanding the socialization of racial minority children and youths. *Harvard Educational Review, 67*(1), 1–39.

Stanton-Salazar, R. D. (2011). A social capital framework for the study of institutional agents and their role in the empowerment of low-status students and youth. *Youth & Society, 43*(3), 1066–1109.

Stephens, N. M., Brannon, T. N., Markus, H. R., & Nelson, J. E. (2015). Feeling at home in college: Fortifying school-relevant selves to reduce social class disparities in higher education. *Social Issues and Policy Review, 9*(1), 1–24.

Stephens, N. M., Fryberg, S. A., Markus, H. R., Johnson, C. S., & Covarrubias, R. (2012). Unseen disadvantage: How American universities' focus on independence undermines the academic performance of first-generation college students. *Journal of personality and social psychology, 102*(6), 1178.

Striplin, J. J. (1999). *Facilitating transfer for first-generation community college students.* Los Angeles, CA: ERIC Clearinghouse for Community Colleges. (ED 430 27).

Terenzini, P. T., Rendon, L. I., Upcraft, M. L., Millar, S. B., Allison, K. W., Gregg, P. L., & Jalomo, R. (1994). The transition to college: Diverse students, diverse stories. *Research in Higher Education, 35*(1), 57–73.

Terenzini, P. T., Pascarella, E. T. (1980). Student/faculty relationships and freshman year educational outcomes: A further investigation. *Journal of College Student Personnel, 21*, 521–528

Terenzini, P. T., & Pascarella, E. T. (1998). Studying college students in the 21st century: Meeting new challenges. *The Review of Higher Education, 21*(2), 151–165.

Thayer, P. (2000). Retention of students from first-generation and low-income backgrounds. *The Journal of the Council for Opportunity in Education.*

Tinto, V. (1975). Dropout from higher education: A theoretical synthesis of recent research. *Review of Educational Research, 45*(1), 89–125.

Tinto, V. (1987). *Leaving college.* Chicago, IL: University of Chicago Press.

Tinto, V. (1998). Colleges as communities: Taking research on student persistence seriously. *The Review of Higher Education, 21*(2), 167–177.

Tinto, V. (2006). Research and practice of student retention: What next?. *Journal of College Student Retention: Research, Theory & Practice, 8*(1), 1–19.

Valadez, J. R., & Duran, R. (2007). Redefining the digital divide: Beyond access to computers and the Internet. *The High School Journal, 90*(3), 31–44.

Walpole, M. (2003). Socioeconomic status and college: How SES affects college experiences and outcomes. *The Review of Higher Education, 27*(1), 45–73.

Warburton, E. C., Bugarin, R., & Nunez, A. M. (2001). *Bridging the gap: Academic preparation and postsecondary success of first-generation students (NCES 2001-153).* Washington, DC: National Center for Education Statistics, U.S. Government Printing Office.

Warschauer, M. (2002). Reconceptualizing the digital divide. *First Monday, 7*(7).

Warschauer, M. (2003). Dissecting the" digital divide": A case study in Egypt. *The information society, 19*(4), 297–304.

Warschauer, M. (2004). *Technology and social inclusion: Rethinking the digital divide.* Cambridge, MA: MIT Press.

Warschauer, M., Knobel, M., & Stone, L. (2004). Technology and equity in schooling: Deconstructing the digital divide. *Educational policy, 18*(4), 562–588.

Wei, L., & Hindman, D. B. (2011). Does the digital divide matter more? Comparing the effects of new media and old media use on the education-based knowledge gap. *Mass Communication and Society, 14*(2), 216–235.

Wohn, D. Y., Ellison, N. B., Khan, M. L., Fewins-Bliss, R., & Gray, R. (2013). The role of social media in shaping first-generation high school students' college aspirations: A social capital lens. *Computers & Education, 63*, 424–436.

CHAPTER 10

THE HABITUS AND TECHNOLOGICAL PRACTICES OF RURAL STUDENTS

A Case Study

Laura Czerniewicz and Cheryl Brown

This paper describes the habitus and technological practices of a South African rural student in his first year at university. This student is one of five self-declared rural students, from a group of 23 first-years in four South African universities, whose access to, and use of, technologies in their learning and everyday lives was investigated in 2011 using a 'digital ethnography' approach. Their digital practices, in the form of their activities in context, were collected through multiple strategies in order to provide a nuanced description of the role of technologies in their lives. The student reported on here came from a school and a community with very little access to information and communication technologies (ICTs). While the adjustment to first year can be challenging for all students, the findings show that this can be especially acute for students from rural backgrounds. The study provides an analysis of one student's negotiation of a range of technologies six to nine months into his first year at university. Earlier theoretical concepts provide a lens for describing his practices through a consideration of his habitus, and access to and use of various forms of capitals in relation to the fields—especially that of higher education—in which he was situated.

INTRODUCTION

In 2011 we undertook a 'digital ethnography' (Murthy, 2008, p. 1) of students' access to and use of technologies in their learning and everyday lives. During the analysis phase it became clear that it would be important to describe the challenges and adjustments students from rural backgrounds face, both dealing with the transition to university generally, and specifically adopting new technologies into their learning lives. This is not a unique local problem: research from other countries has shown that students from rural backgrounds face challenges succeeding in higher education. A study of various factors affecting students' successful completion at the Open University of Sri Lanka, for example, showed that 77% of rural students did not complete their degrees compared with the 44% of urban students who did not do so (Gamaathige & Dissanayake, 1999). Another study, from Australia, identified students from lower socioeconomic backgrounds living in rural areas as a distinct group at greater risk of educational disadvantage (James, 2002).

In South Africa, it has not been possible to obtain exact figures on the percentage of students from rural backgrounds in South African universities. This is made more difficult because students from rural backgrounds are often included either in the general category of 'disadvantaged students' or assumed to be in the group of students on financial aid.

However, it is clear from the literature that rural schools continue to suffer poor, indeed worse, learning conditions on the whole compared with their urban counterparts. Motala, Dieltiens, Carrim, Kgobe, Moyo, and Rembe (2007, p. 52) note that despite improvements in funding equity, many learners, especially in the rural areas, continue to lack access to proper infrastructure and have to manage with limited text books, badly stocked school libraries and poorly trained educators. They also note that the single most powerful recommendation to emerge from the findings of the Ministerial Committee on Rural Education (Department of Education, 2003 in Motala & Dieltiens, 2010) was the need to improve and equalise facilities and resources. In addition, learners in rural areas are less likely to attend school. Lewin (2009) analyzed Demographic and Health Surveys from 25 Sub-Saharan Africa (SSA) countries and found that urban children have about four times more chance of being enrolled in Grade nine than rural children in the data set. Studies of educational exclusion make the same finding—that more out-of-school children are to be found in rural areas than are to be found in urban areas (Motala & Dieltiens, 2010).

There is also evidence that students in rural schools achieve worse results than their urban counterparts. Motala et al. (2007) observe that learners who fared less well in the Department of Education's national assessments of learning achievement for Grades 3 and 6 came from (in descending order) township, farm, rural and remote rural schools. Many Quintile one schools[1] are in rural areas, with the Free State having 64.1% of Quintile one schools in South Africa, of relevance because four of this study's five rural students (including the student reported on in this paper) were from the Free State. The concerns about these schools and their general poor situation have been addressed in a number of ways, one of which is inclusion in the 500+ Dinaledi schools that receive National Treasury grants to improve mathematics and science results.

The statistics reveal that only 7.7% of the learners in Quintile one schools who wrote the National Senior Certificate exams in 2009 passed (Financial & Fiscal Commission, 2011). This means that the demographic and geographical spread of the country's students is not proportionately represented in universities. One of the few studies which consider rural students in universities (Tumbo, Couper, & Hugo, 2009) reviewed lists of undergraduate students admitted from 1999 to 2002 in Health Science faculties, in terms of whether they had rural, town or city postal codes. They found that 59% of the students were from cities, 15% from towns and 26% from rural areas. They concluded that the proportion of rural-origin students in medical studies at that time in South Africa was considerably lower than the national rural population ratio of 46.3% (and of course disproportionate to the number of medical professionals in rural areas).

While there are numerous studies about the transition to university of disadvantaged students (as defined economically), there appear to be few that focus specifically on those from rural areas, and those specifically from poor rural areas. Kapp and Bangeni (2011) describe some of the challenges students from educationally disadvantaged backgrounds (many from rural areas) face in negotiating aspects of academic literacy at university. They show how students encounter an essentially foreign culture at university, and have to reconcile conflicting transitional spaces of their home and university identities. We have found no studies specifically addressing university students from rural backgrounds in terms of their technological literacies and associated issues. Information from a non-governmental organization (NGO) supporting rural students in higher education says that in student feedback reports on their experiences at university "the unfamiliarity of technology comes up again and again as a real challenge for new students" (Glover, personal communication, 2012). This is not surprising given that Internet penetration is only 4.6% in South African rural areas compared to 21.8% in urban areas (Pejovic, Johnson, Zheleva, Belding, Parks, & Van Stam, 2012). Although mobile telephony is decreasing the location-based divide, the disparity between rural and urban youth is still pronounced; while 70% of urban youth over 16 use mobile phones, only 49% of their rural counterparts do so (Beger & Sinha, 2012, United Nations Children's Fund (UNICEF) New York, Division of Communication, Social and Civic Media Section).

In a study on student dropouts, Brits, Hendrich, Van der Walt, and Naidu (2011) review work reported on by the Rural Education and Access Program (REAP) in 2008 and 2010 which states that a rural background may have a negative impact on student success because students from disadvantaged backgrounds are usually underprepared for tertiary education, but at the same time the preparedness of the institutions to accommodate underprepared students is sometimes questionable

(REAP, 2008, p. 6, in Brits et al., 2011). In their colloquium report of 2010, REAP reports that rural students' success is constrained by financial factors including the fact that many cannot afford to go home for their holidays; academic factors which see them confront the fact that although they have passed Grade 12 they lack the necessary competencies required for tertiary study; and socio-cultural factors, including isolation and alienation in a new environment. In this paper we therefore focus on the story of one student, from what, by his account, is a seriously marginalized context. Through a narrative approach we explore how a student from a rural background navigated the use of technologies in his first year at university and we challenge assumptions that are made about the position of technology in the lives of students from disadvantaged backgrounds. We also explore the learning pathways this student followed in acquiring ICT-based skills needed to successfully traverse higher education.

Theoretical Framework

Students' technology practices cannot be described in isolation; these are not neutral activities. We find Bourdieu's (1984, 1986; Bourdieu & Wacquant, 1992) framework a valuable way to describe student practices in the context of the field in which they occur, and in terms of their habitus and the capitals they are able to bring to bear to their practices.

Bourdieu (1984, p. 101) summarizes his framework as "[(*habitus*) (*capital*)] + *field* = *practice*". A field is a distinct social space consisting of interrelated and vertically differentiated positions, a "network, or configuration of objective relations between positions" (Bourdieu & Wacquant, 1992, p. 97). In the case of this study, we understand higher education to be a distinct social field, and we also understand the rural community in which the students live to constitute a different social field.

Bourdieu (1986) describes four forms of capital: economic, social, cultural and symbolic. Economic capital refers to assets either in the form of, or convertible to cash. Social capital is "the sum of the resources, actual or virtual, that accrue to an individual or a group by virtue of possessing a durable network of more or less institutionalized relationships of mutual acquaintance and recognition" (Bourdieu & Wacquant, 1992, p. 119). Cultural capital occurs in three states: embodied, objectified, and institutionalized. Jenkins (2002) explains that embodied cultural capital refers to long-lasting dispositions of the mind and body, expressed commonly as skills, competencies, knowledge and representation of self image. Objectified cultural capital refers to physical objects as "cultural goods which are the trace or realization of theories or critiques of these theories" (Bourdieu, 1986, p. 243, mentions pictures, books, dictionaries, instruments, machines). Institutional cultural capital is the formal recognition of knowledge usually in the form of educational qualifications. Symbolic capital is appropriated when one of the other capitals is converted to prestige, honour, reputation, fame—it is about recognition, value and status.

Habitus has been widely used in education, specifically to consider the way that disadvantaged students mediate their educational experiences. Bourdieu (1977, p. vii) described habitus as "a system of durable, transposable dispositions which functions as the generative basis of structured, objectively unified practices". This complex definition has been explained to mean that a habitus lens provides a way of showing students' "ways of acting, feeling, thinking and being…how [they] carry [their] history, how [they] bring this history into [their] present circumstances, and how [they] then make choices to act in certain ways and then not others" (Maton, 2008, p, 53). This concept has been critiqued by many scholars as being determinist and leaving little room for agency. In a detailed review of these debates, Mills (2008) argues that the generative dimension of the concept can indicate transformation as well as reproduction. Therefore while an individual's habitus—inculcated by everyday experiences within the family, the peer group and the school—disposes actors to do certain things, orienting their actions and inclinations, it does not determine that they do so. As Mills (2008) points out, Bourdieu explained that habitus is a strategy-generating principle enabling agents to cope with unforeseen and ever-changing situations. The point is that there is no such thing as pure agency; but a kind of (limited) agency can be identified. She makes the point that those with a more transformative habitus recognise opportunities for improvisation and act in ways to transform situations. Thus, she observes that what one may experience as incapacitating, another may see as generative of opportunities for self-enhancement or self-renewal.

As this is not a study of agency, but one of practices, the specific issues of agency—both empirical and theoretical—do not receive attention. Rather the focus is on practices, especially the habitus and the ways that various capitals are drawn on and exerted in the fields, particularly that of higher education.

METHODOLOGY

This paper describes the habitus and technological practices of a South African rural student in his first year at university. It arises from a project which used a 'digital ethnography' (Murthy, 2008) approach to study 23 first-year students in four South African universities (Cape Town, Fort Hare, Free State, and Rhodes universities) during 2011. Digital or virtual ethnographies have been used to expand opportunities for data collection beyond the physical (Murthy, 2008) and to enable researchers to better understand people's technological culture (Hine, 2008). We drew on Ethnographic Action Research (EAR) as a methodological framework as it combined participatory techniques and ethnographic approaches into an action research framework (Tacchi, Foth, & Hearn, 2009). Consequently we recruited and trained researchers located at each study site. The on-site researchers were embedded in the context of the student participants thus available and able to establish a relationship with them. This methodology allowed for the dynamic interchange between researcher and participant; it acknowledged that each influences the other—thus the inclusion of Action Research within its framework. The sources of data for the project (and this paper) comprise a range of interviews (transcribed and viewed), video records and transcripts, transcripts of a focus group as well as digital diaries and social media (Facebook) observations.

The student described in this paper is one of five self-declared rural students from a group. When it became evident that the rural students were of particular interest, we realised that we would lose the richness of their narratives if we tried to focus on all of them. We selected Jake (a pseudonym) purposefully. He had engaged consistently throughout the project and provided an extensive mix of interview and digital data. He had also responded by email when clarification on data and findings was sought.

This paper takes the form of a descriptive case study because it seeks to examine a particular phenomenon, *viz.* the assimilation of technology into the learning life of a rural student in a real-life context, namely, the first year of university (Yin, 2003). In examining this particular event in a real-life context we are looking at a bounded system in its own habitat (Stake, 1978). We acknowledge that this is one person's story and consider it important through the insight it offers into the challenges of a marginalized group of students in higher education.

Data analysis was framed by the theoretical lens and the concepts of Bourdieu's (1986) framework—habitus (with indicators of family background, schooling, and reported competence), economic capital, cultural capital (specifically embodied and objectified), social capital and field.

Transcribed interview evidence for all participants formed the basis of a content analysis that developed four coding matrices based on the research framework. The first three were for the past, current and future use of ICT by students. The fourth defined their use of varied aspects of social media. These were combined into a highly structured coding matrix using NVivo software which was used to code all the students' interviews. In addition, further analysis of all the data sources was undertaken for selected students' cases. In the case of Jake, we also sought clarity on further issues via email, and gave him the opportunity to read the draft of this paper. Whilst we sought informed consent from all students for the research and guaranteed their anonymity in publication, it was important that an individual student view our academic interpretation of his story.

Jake's Story

Habitus is a way of describing how, "amongst other things [students], carry [their] history, how [they] bring this history into [their] present circumstances" (Maton, 2008, p. 53). Therefore Jake's personal background is relevant to understanding his later digital practices.

Background

Jake, aged 19, is a first-year Humanities student, the first person in his family to go to university. He is trilingual, speaking seSotho as a home language but also fluent in isiZulu and English. His mother, who worked in town as a factory worker, had died when he was 12 years old and he described his father as a self-employed tiler. His father had two wives and four children, all of whom Jake considered his siblings (he spoke of his sibling on his mother's side, and his siblings on his father's side). Neither of his parents successfully completed their schooling; his mother left before the end of her school years and his father failed the final school exams. Jake is the first person in his family to successfully finish school and be accepted for university.

Jake attended primary and high school in small rural schools, the primary school having fewer than 100 pupils, the high school under 500 pupils. His high school was one of the afore-mentioned Dinaledi schools focusing on mathematics and science. Because of the additional grant these schools received, his school did have some technology, but as noted below, he did not have access to it.

Economic Capital and Access to Technology

Jake's access to the types of capital which he could draw on in the field of higher education was limited. In terms of economic capital, there was a substantial shift when he got a university bursary that paid for his laptop and smartphone. Before he came to university he had had few encounters with technology. His school had two computers in the library but he had never touched them. He got his first cell phone when he was in Grade 10, and until he came to university he thought a cell phone was the most important item of technology to have.

He says that when people in his village need to access the Internet from a computer they have to travel to an Internet cafe and that "it is costly". He makes a similar observation about accessing the Internet from cell phones when he notes that, "Most of my peers had cell phones that had internet but they couldn't use them to surf the net due to coverage issues and costs". He says he now uses the Internet more than his friends at home because,

> when I get to the internet, already having like paid for the data services and stuff, I get for free. So to them it is still like they have to budget for the whole thing, so I think it is more accessible to me than it is to them.

It is through the economic capital Jake acquired when winning the bursary that he gained access to what he describes as the "individualism" of his own laptop. He says that the university does have computer labs but that he prefers having his own machine, which he points out he uses differently. Having the luxury of an individual machine means that "there are no monitors" and that while in labs, "there are other students and you have to be silent". He notes that, "with your own computer you can angle it the way you want and watch whatever you like".

Having the money to buy a laptop and phone also meant that he gave away other devices such as flash drives and memory sticks to others "at home, for music and stuff".

If considered in terms of his university context, his economic capital or resources are mixed. On the one hand Jake's computer at six months old is considered relatively old, and he describes it as, "problematic. Full of *matata,* problems. This Acer of mine is now becoming too old". But Jake is mindful of the fact that he is on a bursary and says that he would never ask his family for money to buy technology as he cannot justify it to them; he points out that they would not understand what he was asking for. He also adds that he still has credit on his bursary, so this is another reason not to ask his parents.

While the bursary has literally provided Jake with economic resources he did not previously have, he continues to operate in an environment where technology practices are shaped by resource constraints. Thus he does all his printing in the university labs, as he does not have a printer. And he notes in one of his Facebook status updates that: It ws a shock @ first 2 notice dat many of my frnds r no longer updating anythng on fb, bt then again I got reminded of da fact dat skuls r closed. Whch minz no more easy access 2 free computers on campus. Only dose wit smartphnes r beatin da recession. # shame #.

Jake is aware of the privilege his financial position offers him and that without access to centralised university resources such as computer labs, his fellow students are cut off from their online social networks.

Social Capital

Acquiring a computer has given Jake access to social capital and changed the dynamics of his relationships.

He is able to help his room-mate as he is the one who has the computer. Jake shares his laptop with his room-mate when he needs the Internet, otherwise he would have to travel a distance to gain access. Jake says:

> I'm logging in…to send an SMS. My room-mate's mother. Ja, [he] is looking for money, money. …My room-mate wants to send an SMS, let me just login. My room-mate is going to send an SMS to his Mama. Apparently, he's looking for something like money, whatsoever, so I am going to make space for him.

He has gained new networks, and he calls on them for help if he has problems with his computer, particularly his network of neighbors in the hostels. He says that he has never asked for help in the computer labs; even though there are plenty of people there. He says it takes too long for anyone to respond if he raises his hand, observing that "they don't even see you".

His online networks have changed substantially in the six months that he has had the laptop and the Blackberry, as he did not have a computer before (and nor did anyone in his family) and his previous cell phone did not have easy Internet connectivity.

He uses Facebook on the computer regularly (at least every hour) and both Facebook and Mxit on the Blackberry. He has between 350 and 400 Facebook "friends", of whom he says only 15 are friends known to him personally. The networks are through his church, his home community and through his connections in another province. He is very active on lists and groups, many of them related to his church and other religious groupings. He says that Mxit is largely for a small group of his church friends.

Cultural Capital

Jake is aware that before he came to university he had limited cultural capital in terms of computer literacy, and he observes that although there were in fact two computers at the school he did not use them because he did not know how "to compute or surf the net. In our school we didn't have computer literacy skills."

He notes that "the technological infrastructure in my home comprised only a radio". He says, "In our community no one has a computer" and he adds that the only technology that matters in his home village is "music systems". He does not link the lack of technology to lack of finance or economic capital, but interestingly, rather to cultural capital. Thus, he says, "In my community technology is not that important as people lack the skills to use it."

"Before I came to varsity technology was important as I understood its vitality (*sic*) in the Information Economy we are living in. I always dreamt of owning a laptop." As soon as he got to university, he did a first-year basic computer literacy course for six months. He said these months were tough and that he failed the first test and needed help with the practicals. He continues to use the computers in the lab, as they have software programs that he does not have on his laptop—he is struggling to finish a PowerPoint and Excel installation, for example. Jake is expressive when he explains that:

> My rural background was very challenging upon my arrival at the UFS. The computer literacy module, BRS 111, was like rocket science to me. To top it up, lecturers had started making use of Blackboard for notifications. It was a real challenge. Typing assignments in accordance with the required formats was even more threatening. Universities are congested with computers and it is unwittingly assumed that all students can use those computers beneficially; which is not the case. Rural students feel estranged and depressed by these technologies. Setting appointments via email, checking emails from the varsity, and doing research for assignments are scary activities to rural students.

He is proud of what he has achieved, and says, "it is becoming easier and easier as I get familiar more with the technology…. Now I'm like accustomed to everything and I know the functions, ja. Now, I love it."

Objectified Cultural Capital

The importance of both the computer and the Blackberry in Jake's life is captured by two of his Facebook statuses. The first reads: "Imagine life without this tiny device called a Blackberry. Imagine life without this cute portable thing called a laptop...jerrrr.....tasteless." The second reads: "nna to be honest these two r rulin my world..of course dey come after my beautiful galfriend."

For Jake, his computer is a valuable artefact: "I don't want to play games on my laptop, as the heavy use of keys may damage my keyboard." It is an important part of his life now, as a student. He says,

> The thing is that at Varsity level you can't survive if you do not have your own laptop. Cos' like, truly speaking, we have computer labs, so before I bought my computer, I still had access to computers, but I just felt that I needed my own." He estimates that he spends 80 hours on his laptop a week, guessing that over half of that is on "academics" with the rest being entertainment, particularly music. Of his six-month journey, Jake says, "I know my way around the computer everything is satisfactory, it is how we live now." For Jake, the laptop is about his studies and his music: "So, my laptop, it has to do its job; assist me academically and entertain me. I can't just put it away and not play on it, then it's not doing its job.

Yet, while Jake prizes his computer, he believes his cell phone is indispensable. He says, "I can't imagine life without my cell phone"—it is always with him and he never switches it off. He would find being in an area without cell phone reception problematic. He says that he will answer texts at the dinner table and that his concession to university lectures is to log out of Facebook and Mxit on the phone, although he says he still receives notifications even when he is logged out, and that he logs on again as soon as the lecture is finished.

He says the phone is really useful when he is waiting in a queue, and that having the phone on the four-hour taxi journey home makes the journey seem faster.

He is honest that having a cell phone is about style and reputation. He says that his previous phone was "an old Nokia, from my father. The phone had no camera, couldn't play MP3s, and was worn out." He said at school, "you want to have a cell phone, you don't care about emails and stuff. You want to fit in. It is the trends." When he came to university and won the bursary he decided to buy a new phone. He observes that his new phone is really easy to use, but notes that ease-of-use is not the reason he bought it. Rather he says: "I actually switched to it, because it was popular. So, I got to learn about its functionalities after I had bought it."

After a few months of having the phone Jake became worried about the effect it was having in his studies. He says:

> If you want to access Mxit, Facebook and all social networking sites, you actually don't like concentrate on your work. One minute you study, there comes a notification; 'Someone posted on your wall', 'Someone has sent you an email' or something like that and you have to go and check it out. So, I decided to cut off my Blackberry Internet Service (BIS) and I did not renew it for this month. I will only like renew it when we go to recess, because of these like tests and stuff, ja. At the moment I only access Facebook when I am in my room at night.

Digital Practices

It is evident from his self-recording that Jake uses his two devices in a variety of ways, and that he multitasks. He moves between devices, for example when he writes on a friend's Facebook page, "Hey monna inbox me ur mxit..." in order to have a private conversation. It is also clear that he uses social media for both academic and personal purposes, an indication of the kind of blurring that happens with both software and hardware. While most of his Facebook statuses (and comments on his friends' statuses) are personal (including many quotes from popular songs and religious references) they also include affective comments and reflections including:

- Study hard n al shall follow.
- Stop whining n study.
- Good luck 4 da second semester.
- I so wish textbooks were designed lyk magazines. Ten years frm nw u probably wud stl rememba da story of Kelly Khumalo in Drum Magazine, whch minz evn chapters wud last 4eva in one's mynd.
- [names a friend] I'm studyin nw. Procrastination is a rapist of tym.

He is on video record using his Blackberry as a modem and emailing his work to other students with whom he will be working in the lab. In one clip he is using Mxit on his Blackberry, Facebook on his computer, and is also playing music. He thinks about technology and notes in one of his Facebook statuses that he does not like Twitter:

> I jst cnt connect with Twitter. Its so shortcut n complex. @ Jake. @ Kido...@ Nomvie...such titles n da scribblings confuse me.

He also says, "So, Twitter is more competitive than Facebook, Facebook is how you feel."

It is interesting that Jake perceives his laptop to be part of the academic world he has entered, an aspirational part of the information economy, and also how closely he links expertise, skills and training with the computer as specific type of technology.

Field

Jake sees the challenges he has experienced with technology as part of the wider sweep of challenges that he has had to deal with in the transition to university. He is aware of the dramatic effect that this can have and he observes that not only for him, but also for others, these transitions can be a shock:

Since campuses appear to be sophisticated, rural students prefer to abstain from exploring them with the fear of their rural mentality being exposed. Apart from technology, lecturers make it tough for rural students because these lecturers are used to Model C schools and the way those schools teach. As a result, their tone suits private school graduates. Their language and examples are in line with these urban students. No one seems to understand township or homeland students. His habitus can be described as transformative both in terms of his experiences at university and at home. In both cases there have been changes. His relationship with his class mates has changed as it has with his home community. He says that while at university he knows less than his friends, at home he is "the mastermind". Being ICT literate gives him a kind of freedom, "And I think that I am the only computer literate one in my family. So, I am free to do anything I want, because they don't understand what it is." Fellow students ask if his family know what Facebook is and laugh; Jake replies, "They do not understand."

His friends, family, relatives and teachers back home think he is "clever with technology" and they ask his advice about things, for example, about "how a web-cam functions". He is the first person in his family to own a laptop.

He comments that while he is connected to his university lecturers and emails them, he is not in contact with teachers from his school as they are "not that much computer literate". For example he says that they hand work to the school clerks to type for them, "to do reports and stuff". With regards to his friends at home, he says he largely uses Short Message Service (SMS), that he cannot Blackberry Messenger (BBM) them as many of them do not own Blackberry cell phones. When he returns from a short break at home he notes, "At home I did not use any technology, at all –

Kwala

There is one area of continuity, that of music. He had noted that the only technology that matters in his home village is "music systems". Music continues to be very important to him. "The first thing I do always is play music [on my computer]; …But everything I do, I do to music." This is the one aspect of his new computer that he values highly, demonstrating in the recorded videos, how his Virtual disc jockey (DJ) system works. He is clearly proud of the expertise that he has gained in this arena.

CONCLUSION

While this paper focuses on students' technological habitus, it also demonstrates that technological literacies are interwoven with other literacies as only one component in a dense skein of experiences and adjustments for first year students.

Bourdieu (1986) provides a useful conceptual lens for describing how one student uses and engages with technology, as well as framing the way his background shapes his response to his new environment. The findings highlight an example of a transformative habitus—as Jake himself says in relation to his background: "I'm not sorry of myself, despite my regrettable background. Actually, it's that background which keeps reminding me that I have to work even harder to stay competitive." Jake's ability to develop symbolic capital through the motivation his background has elicited in him thus provides an additional thread that runs through his story.

Jake's story serves as a stark reminder that for students from disadvantaged backgrounds—in this case from a rural background—access to success in higher education is more than just entrance to university. It is about on-going and continuing access to various forms of capital such as financial assistance and the establishment of new support networks. It is about the disjuncture between background and institutional culture, and the consequent reforming of individual habitus in the light of the imperatives of the higher education terrain. In negotiating this path, technology has played both an inhibiting and enabling role for Jake. While it was initially yet another challenge to navigate, it proved to be an enabler to building his confidence, establishing new connections and providing new opportunities for him to pursue his passions. The multiple forms of data provided by the 'digital ethnography' methodology of this study afforded a lens to Jake's technology practices and provided a view of his strategies for success.

NOTE

1. The poorest schools are included in Quintile one and the least poor in Quintile five. Schools are classified first by a national poverty table, prepared by the Treasury, that determines the poverty ranking of areas. It is based on data from the national census including income levels, dependency ratios and literacy rates in the area. Secondly the provinces then rank schools from Quintile one to five, according to the catchment area of the school. Each national quintile contains 20% of all learners, with Quintile one representing the poorest 20% and Quintile five the wealthiest 20%. However, provincial inequalities mean that these quintiles are unevenly distributed across provinces.

REFERENCES

Beger, G., & Sinha, A. (2012). *South African mobile generation: Study on South African young people on mobiles. Digital citizenship society.* New York, NY: United Nations Children's Fund (UNICEF), Division of Communication, Social and Civic Media Section, Retrieved 18 June 2012 from http://www.unicef.org/southafrica/SAF_resources_mobilegeneration.pdf

Bourdieu, P. (1977). *Outline of a theory of practice.* Cambridge, UK: Cambridge University Press.

Bourdieu, P. (1984). *Distinction: A social critique of the judgement of taste.* Boston, MA: Harvard University Press.

Bourdieu, P. (1986). The forms of capital. In J. Richardson (Ed.), *Handbook of theory and research for the sociology of education.* New York, NY: Greenwood.

Bourdieu, P., & Wacquant L. (1992). *An invitation to reflexive sociology.* Chicago, IL: University of Chicago Press.

Brits, H., Hendrich, U., Van der Walt, M., & Naidu, Y. (2011). *Student dropout at the Vaal University of Technology, a case study.* Centre for Academic Development, South Africa: Vaal University of Technology.

Financial & Fiscal Commission. (2011). *2012/13 Submission for the division of revenue.* Pretoria, South Africa: Midrand & Cape Town: Financial & Fiscal Commission. Available at http://www.ffc.co.za/docs/submissions/dor/2012/FFC%20SDOR%20 -%20Approval.pd f. Accessed 12 December 2012.

Gamaathige, A., & Dissanayake, S. (1999). Comparison of some background characteristics of students who have completed and not completed the Foundation Programme in Social Studies. *Open University of Sri Lanka Journal, 2,* 65–79. doi: 10.4038/ouslj.v2i0.364

Hine, C. (2008). Virtual ethnography. In L. Given (Ed.). *The SAGE encyclopedia of qualitative research methods.* Thousand Oaks, CA: SAGE Publications.

James, R. (2002). *Socioeconomic background and higher education participation: An analysis of school students' aspirations and expectations.* Melbourne, Australia: Commonwealth Department of Education Science & Training. Available at http://www.voced.edu.au/content/ngv2433. Accessed 18 June 2013.

Jenkins, R. (2002). *Pierre Bourdieu* (Rev. ed). New York, NY: Routledge.

Kapp, R., & Bangeni, B. (2011). A longitudinal study of students' negotiation of language, literacy and identity. *Southern African Linguistics and Applied Language Studies, 29*(2), 197–208. doi: 10.2989/16073614.2011.633366

Lewin, K. (2009). Access to education in sub-Saharan Africa: patterns, problems and possibilities. *Comparative Education, 45*(2):151–174. doi: 10.1080/03050060902920518

Maton, K. (2008). Habitus. In M Grenfell (Ed.). *Pierre Bourdieu: Key Concepts.* London, UK: Acumen.

Mills, C. (2008). Reproduction and transformation of inequalities in schooling: The transformative potential of the theoretical constructs of Bourdieu. *British Journal of Sociology of Education, 29*(1), 7–89. doi: 10.1080/01425690701737481

Motala, S., & Dieltiens, V. (2010). *Educational access in South Africa, country research summary.* UK: Centre for International Development, University of Sussex. Available at http://www.create-rpc.org/pdf_documents/South_Africa_Country_Research_Summary.pdf. Accessed 12 December 2012.

Motala, S., Dieltiens, V., Carrim, N., Kgobe, P., Moyo, G., & Rembe, S. (2007). *Educational access in South Africa: Country Analytic Review.* Johannesburg, South Africa: The Education Policy Unit at the University of the Witwatersrand.

Murthy D 2008. Digital ethnography: An examination of the use of new technologies for social research. *Sociology, 42*(5):837–855. doi: 10.1177/0038038508094565

Pejovic, V., Johnson, D. L., Zheleva, M., Belding, E., Parks, L., & Van Stam, G. (2012). The bandwidth divide: obstacles to efficient broadband adoption in rural Sub-Saharan Africa. *International Journal of Communication, 6,* 2467–2491. Retrieved 7 January 2014 at http://www.cs.ucsb.edu/~ebelding/txt/Pejovic2012IJOC.pdf

Stake, R. E. (1978). The case study method in social inquiry, *Educational Researcher, 7*(2):5–8. doi: 10.3102/0013189X007002005

Tacchi, J., Foth, M., & Hearn, G. (2009). Action research practices and media for development. *International Journal of Education and Development using Information and Communication Technology, 5*(2), 32–48.

Tumbo, J. M., Couper, I. D., & Hugo, J. F. M. (2009). Rural-origin health science students at South African universities. *South African Medical Journal, 99*(1), 54–56.

Yin, R. K. (2003). *Case study research: Design and methods* (3rd ed.). Thousand Oaks, CA: Sage.

PART III

GLOBAL RESEARCH AND DEVELOPMENT IN TECHNOLOGY

CHAPTER 11

THE DIGITAL DIVIDE IN SCIENTIFIC DEVELOPMENT AND RESEARCH

The Case of the Arab World

Hamoud Salhi

A contributing factor to the deficiency in human development in the Arab world is attributed to the increasing gap in the scientific knowledge that exists between the Arab states and developed world. In its 2012 annual report, the World Bank concluded that the Arab world has a much lower knowledge indices value than most of the regions in the world, and that its average performance only exceeds that of Africa and South Asia. This chapter examines the effectiveness of the ICT strategies adopted by Arab national governments to reduce the digital divide. I will observe the activities undertaken by Arab governments in the area of scientific knowledge building as a means to reduce the digital gap in the Arab World. It is hypothesized that the higher a country is spending on research, development, innovation, and ICT, the higher its researchers' scientific production will be. I anticipate this production to be low considering the amount spent by Arab governments on research and development. This sets us to conclude that the digital divide in the Arab World has the potential to grow even wider given the lack of emphasis by Arab government leaders on innovation and the preparation of a scientific generation to lead this region in the Digital Age.

Recent studies (Al-Emran, Elsherif, & Shaalan, 2016; Bezuidenhout, Leonelli, Kelly, & Rappert, 2017; Klischewshi, 2014) on the digital divide in the Arab World have concentrated on the degree of penetration of the various Information and Communication Technology (ICT) devices in the region, the total number of users, and how suitable the regulations and laws for the expansion of ICT. Countries that fare well in these categories are expected to have a narrower digital divide, much closer to that of advanced countries. Thus, the digital divide is treated as a measure to detect how advanced a country in its integration in the global technological world. Inherent is the assumption that digitalization is the ultimate goal of every nation in the world, and that not only is digitalization inevitable but also beneficial (Hillbert, 2001; Pearce & Rice, 2013). This study challenges these assumptions. This study sees merit in using the digital divide as an analytical tool; however, the degree of digitalization can lead to erroneous conclusions, falsely projecting that the acquisition of the highest number of ICT devices and having a large percentage of Internet users, coupled with having the most efficient regulations (read bureaucracy), are sound indicators to label a country digitally advanced. Based on the Arab World's case, the author argues that digital divide measures fail to account for the dependency mechanisms that are created when ICT is introduced in the region. The Arab World's heavy reliance on importing technologies and on foreign expertise to implement and sustain them risk the development of a dependent ICT infrastructure. It may likely increase the digital divide as the region finds itself unable to keep up with the financial costs of ICT or draws resources away from other national priorities.

Accordingly, the author examines the effectiveness of the ICT strategies adopted by Arab national governments to reduce the digital divide. I will observe the activities undertaken by Arab governments in the area of scientific knowledge building as a means to reduce the digital gap in the Arab World. It is hypothesized that the higher a country's spending on research, development, innovation, and ICT, the higher its researchers' scientific production will be. I anticipate this production to be low considering the amount spent by Arab governments on research and development which is less than 1 percent of the total GDP, according to the UNESCO report 2015. This sets us to conclude that the digital divide in the Arab World has the potential to grow even wider given the lack of emphasis by Arab government leaders on innovation and the preparation of a scientific generation to lead this region in the Digital Age.

The Arab World Integration in the Global Network

Just when technology was in its highest peak in the mid-1990s, and when dot. coms, start-up companies, and a new class of entrepreneurs were making their way to the top in the Western World, Bill Gates predicted the Middle East would be the last to see the benefits of this technological revolution. At the time, callers to a national radio show in the United Arab Emirates (UAE) were condemning the spread of ICT in the Middle East and Northern Africa. They branded Internet connectivity as a real threat to their societal fabric and Islamic upbringing. Further, religious and secular scholars projected a narrative that described ICT as evil, a manufactured ploy designed, once again, to promote Western civilization and to undercut their superior Islamic way of life. Stopping this threat was a trending topic in the media and public discourse.

Today, technology has slowly inched into nearly every public and private sphere of the Arab World. A massive revolution in communication is rapidly transforming Arab society, its culture, politics, and economy (Loch et al., 2003). ICT is proving a liberating tool for women's emancipation, their engagement in economic activities, and owning businesses outside patriarchal control (Shamaileh, 2016). Politically, more individuals and groups are exposing the malaise of authoritarianism and its corrosive impact in unprecedented ways. Political bloggers and commentators are more visible now than a few years ago, when writing letters to newspapers editors was the only means for readers to express themselves. To counter this change, governments have been more alert to responding to threats with censorship, harassment, and unwarranted arrests. But that has not stopped individuals and human rights associations from overcoming governmental barriers. (McGarty et al., 2013)

The Political Impediment

In his commentary at the Brooking Institute on the state of building a knowledge society in the Arab World, Rami Khouri (Brookings Institution, 2008), commentator and editor-at-large of the Lebanese newspaper Daily Star, deplored the notion that change is not happening in the Arab World and that the recommendations brought forth by the report's writers were a magical formula for a better tomorrow. He stated: "We agree that we need change, we know what the problem is, we know what we want to reach, where we want to move to. It's all outlined very well." (Brookings Institution, 2008) Then he added:

> What we don't know is how do we start that process? How do we make the changes in the exercise of power that need to be made to get the kind of results that we want? We don't know how this will happen in the Arab World, we don't know if it'll be a Lech Walesa model, it'll be a Gorbachev model, al-Gaddafi model, a change from the top or Iranian style revolution model." (p. 28)

Khouri's point is that change is "already happening" in the Arab World, but not in the "efficacy and quality of the government systems that bring about changes that we need here for a society of knowledge, and therefore for prosperity and economic change."(Brookings Institution, 2008, p. 30). To Khouri, the real change that would bring knowledge building must occur at be at the top of the political structure: "a reconfiguration of the state that would undercut the political constraints that are inhibiting the real growth of a knowledgeable Arab society." (Brookings Institution, 2008, p. 31). This is best reflected in how Arab governments have come to approach the Arab World's transition to the technological phase.

The Strategy and Digital Divide

A key focus of several Arab states is creating an infrastructure suitable for developing broadband technologies, seen as the driving force that would help them narrow the digital divide that separates them from the advanced countries and modernity. Arab countries expect such technologies to stimulate the economy, create job opportunities for their populations, and achieve prosperity. Scholars and IT experts further predict that the development of broadband Internet will increase the productivity of "multiple sectors of the economy (energy, water, industrial production, and so on)" in ways similar to the impact of the printing press, steam engine, and electricity during industrialization (Gelvanovska, Rogy, & Rossotto, 2014).

Accordingly, the Arab states have set in motion several projects and international initiatives to lay the foundation for broadband technologies. The UAE, Qatar, Saudi Arabia, Algeria, Tunisia, Egypt, and Lebanon have slowly implemented several developmental plans for broadband networks and satellite systems to broaden and enhance Internet connectivity throughout the Arab World. Additionally, the Arab League, Algeria, Tunisia, Bahrain, and Qatar have hosted international conferences, forums, and workshops to gain the expertise in ICT. These meetings provided opportunities for the Arab countries to create their own broadband strategies to increase Internet penetration and sustain it efficiently (Gelvanovska, 2014, pp. 35–45)

There is a consensus in the literature that the Arab World is better connected to ICT than it was years ago, but the region is still in the periphery, lagging behind in many other areas, including digital content, infrastructure, human resource development, skill sets, and bureaucratic efficiency (Dutta & Coury, 2003; Murphy, 2009). According to Murphy, the region is fast growing in Internet connectivity and is progressing at a rate much faster than what industry analysts predicted it would be. (Murphy, 2009).

The ICT development index shows four Arab countries of the Gulf region (Bahrain, the UAE, Qatar, and Saudi Arabia) ranked among the highest ICT penetrated countries in the world in 2016 and 2017 (see Table 11.1). Comparing the Arab states of the Gulf region with 22 other countries from Asia and the rest of the Arab World, Kaba and Said (2013) found that the Gulf region had narrowed the digital divide due in part to its adequate infrastructure and the Gulf governments' commitment to ICT to bridging the digital gap. Furthermore, the authors concluded their comparison by affirming that a well-established infrastructure is a strong contributor to the reduction of the digital divide (Kaba & Said, 2013, pp. 358–364).

Real obstacles still exist in the Arab World and continue to undermine efforts to bridge the digital divide. The literature points to the relative high cost of Internet connectivity in most Arab countries, causing its Internet activities to lag behind the rest of the world. Additionally, a significant number of Arab states do not have the resources to catch up with the speed of change in technological innova-

TABLE 11.1. ICT Development Index 2017 and 2016

	Rank	IDI 2017 Value*		IDI 2016 Value*
Bahrain	31	7.60	30	7.46
UAE	40	7.21	34	7.18
Qatar	39	7.21	36	7.12
SaudicArabia	54	6.67	45	6.87
Oman	62	6.43	64	6.14
Lebanon	64	6.30	65	6.09
Jordan	70	6.00	66	5.97
Kuwait	71	5.98	70	5.75
Tunisia	99	4.82	95	4.70
Morocco	100	4.77	98	4.57
Algeria	102	4.67	106	4.32
Egypt	103	4.63	104	4.44
Syria	126	3.34	124	3.32
Sudan	145	2.55	141	2.56
Mauritania	151	2.26	152	2.08
Comparative Data				
Iceland	1	8.98	2	8.78
Korea (Rep.)	2	8.85	1	8.80
France	15	8.24	17	8.05
USA	16	8.18	15	8.13

Source: ICT Development Index, 2017 http://www.itu.int/net4/ITU-D/idi/2017/index.html

* The ICT Developmental Index (IDI) is the sum of 11 indicators added together to form a benchmark measure used to identify a country's performance overtime and at three ICT stages: readiness, intensity, and impact. https://www.itu.int/en/ITU-D/Statistics/Pages/publications/mis2017/methodology.aspx

tion or have more pressing priorities to allocate resources. This is compounded by the fact that the Arab World strategy of digitalization relies heavily on import substitution more than on local research and development. (Alrawabdeh, 2009; Alghadheeb & Almeqren, 2014).

Additionally, the ICT Development Index shows a wide disparity among Arab states in Internet connectivity. While Bahrain, Lebanon, Oman, and the UAE place in the top 50 countries of the ICT index, the rest of the Arab World countries have the lowest rankings, and are among the least technologically connected in the world, scoring above 100. The literature attributes this divide to disparities in income, leadership in government, and population resistance to ICT services (Dutta & Coury, 2003). Countries with low incomes, like Mauritania, Sudan, and Egypt, have less financial capability to develop a local ICT infrastructure. Consequently, they became reliant on outside expertise for the bare minimum of technological connectivity and service.

Another contributing factor to the growing digital divide in the Arab World is related to the lower status of the Arabic language. (Murphy, 2009, p. 145) The Arab countries have made modest efforts to digitalize Arabic content. According to many sources, Arabic content is only 0.162 percent of the world's digital content, and of the total number of articles published by Wikipedia only 1.34 percent are in Arabic. Further, only 42 percent of all the webpages published within the broader region of the Middle East and Northern Africa are in Arabic. Last but not least, the MENA region hosts only 0.198 percent of the total world number of webpages.

To date, the Arab states have tried to address these issues, but with limited success. For example, the Egyptian Ministry of Communications and Information Technology have led efforts to digitally document the heritage of the Arab World while Bahrain, along with the United Nations Economic and Social Commission for Western Asia (UNESCWA) and the Egypt Information Technology Institute, organized a competition on innovation in 2013.

The Strategy and the State of Research and Development

Building a knowledgeable scientific society is key to the progress to any country. It lays the foundation for the establishment of key industries and provides the country with the tools necessary to meet its developmental challenges. This is very crucial in today's technological world. Acquiring information and knowing how to access and store it are increasingly becoming requirements of our daily activities.

Examining the state of research and development in the Arab World can help us evaluate its capability to narrow the digital divide. We concentrate first on the scientific production outcomes and patent registrations to determine if scientific production is occurring in the Arab World and at what level. Previous research and reports (UNESCO, 2015; UNDP, 2011) show that among the contributing factors to the deficiency in human development in the Arab World is the increasing gap in the scientific knowledge that exists between the Arab states and developed world. In its 2012 annual report, the World Bank concluded that the Arab World has a much lower knowledge indices value than most of the regions in the world, and that its average performance only exceeds that of Africa and South Asia. The report depicted the gap in the areas of scientific innovation, research and development, and the share of total research publications in the world, as severe and dangerous. A similar report drawn by the United Nations Development Program in 2014 called the scientific environment in the Arab World alarming. Further, the report raised concerns over the current policies, questioning their effectiveness to alter the existing path of human development in the region.

Market-Led research

The market-led research reflects the strategic developmental priorities of the Arab leaders (Al-Emran et al, 2016; Loch, Straub, & Kamel, 2003; Murphy, 2009, Pearce & Rice, 2013). As the national strategy emphasizes ICT connectivity, access and penetration, research in the Arab World has been directed at exploring and finding the suitable environment to better navigate products and customer service (Bezuidenhout, Leonelli, Kelly, & Rappert, 2017). After years of delays, several countries are now entrenched in the technological sphere, albeit partially. Gulf countries in particular are exhibiting strong signs that they are on their way to becoming the tech-hubs of the world. The UAE is preparing itself to be among the most advanced telecommunication networks, attracting Western companies for launching the latest Iphone model to the region and Asian countries. At the same time, the UAE is venturing into space research and exploration with advanced countries, including the U.S., and exploring human life in outer space, while Egyptian companies are pursuing ventures in Northern Africa and Korea.

Similarly, several Arab states are transitioning their governmental institutions into e-government modes and promoting e-commerce. As a result, a plethora of start-up companies and dot.com ventures are flourishing throughout the region, particularly in Egypt, Saudi Arabia, the UAE, and Oman, to name a few. This has created new interest for small businesses, which necessitated new electronic regulations. As such, the majority of Arab states are signatories of international treaties regulating ICT globally. Such developments have led to a plethora of studies on e-commerce, e-government, and cyber security threats, along with significant technical literature on software, addressing support, maintenance, and applicability to the region.

Within this marketability framework, data show that some Arab countries have had success creating a market environment for future success. According to the 2016 Global Information Technology Report that ranked countries' capacity to innovate there are only four countries from the Arab World in the top 50 among 139 countries in the world. Countries such as Qatar (12), UAE (28), Lebanon (45), and Jordan (47) have a better score than the rest of the Arab World for their potential capability to adopt, process, and implement technologies at low cost and high productivity. Most of other Arab countries ranked among the lowest in the world, including Algeria (126), Egypt (132), and Mauritania (139). This noticeable difference among the Arab states can be attributed to the maturity of market institutions, as in Jordan and Lebanon, and market efficiency, as in the UAE. Both Qatar and the UAE have highly developed ICT laws (ranked third and fourth, respectively), and while Jordan (44) and Lebanon (47) do not fare as well, they have a long history with private ventures and an evolving capitalist market. The rest of the Arab World's low ranking appear to be consistent with their overall performance of ICT laws, which are listed as not developed.

Another indication of the degree of support Arab government's show for bridging the digital divide is the extent to which the decisions they make are in support
of technological innovation, albeit connected only to marketability. As we have
seen, there are significant differences in the assessment of the Arab governments'
procurements of advanced technological products. With the exception of Oman
(ranked 43) and Kuwait (101), the GCC countries are well above the rest of the
world. Qatar and the UAE ranked first and second, respectively, for their decision
to acquire the most advanced technological products. The rest of the Arab World
did not follow course, with a large number of states scoring very poorly—Tunisia,
Lebanon, and Mauritania among the lowest supporters of advanced technologies.
However, these differences can be interpreted because of each state's capability to
spend on expensive technological products as much as it can be due to differing
strategic priorities in face of other more pressing needs (The Global Information
Technology Report, 2016, p. 220)

Still, as impressive as this may be, it appears as if the Gulf States are trying
to buy themselves out of the digital divide. Buying the best and most advanced
technology from the market has yet to usher in homegrown technologies. Japan
and China are notorious for their "copycat" products. But their approach (steal it,
copy it, and rebuild it to make it even better) has worked for them. In the Gulf,
the mindset is different: if we can afford it, we buy it. It is clear that technology
is defining the Gulf and not the other way around. For example, take the medical
field in the Gulf countries. Several hospitals have advanced technologies for testing and cure. In addition, many have partnerships with hospitals in the U.S., and
have collaborated with American clinics in medical and administrative projects.
As these hospitals praise themselves for providing the best quality service, they
see partnerships with prestigious American medical institutions as a way of not
only delivering on that promise but also promoting their facilities.

But such partnerships do not seem to include shared research and collaboration. For example, the U.S.-based Mayo Clinic partnership with the American
Hospital in Dubai has included workshops on administrative leadership, and organized medical activities tend to highlight the meetings with the political leadership of the country more than the actual delivery of scientific outcomes. What this
means is that much of the UAE's efforts to narrow the digital divide in the medical
field will depend on its purchasing power. But there is more to this point.

There is potential benefit for research. In the long-term, it can reduce cost and
generate more profit. For example, it took $2.5 billion and 10 years of scientific
research on the Human Genome Project (HGP) for the International Consortium
of Scientists to deliver a draft of a DNA blueprint for human life in 2013. The
effect of this discovery has been extraordinary: it ushered in the era of "personalized, or precision, medicine." (Heusel & Richards, 2017/2018, p. 52). With time,
this revolutionary discovery led more doctors to using it to understand disease and
cancer, and the cost for while-genome sequencing analysis reduced to $600 and
for a shorter time (Heusel & Richards, 2017/2018, p, 52).

At the same time the HGP launched, the Dubai Government established the Center for Arab Genomic Studies with the mission to improve the quality of human life by preventing genetic disorders in the Arab World using the latest discoveries. After 15 years, the Center's achievements appear to be in areas other than discovery or research. It amounts to holding a series of conferences and a number of published papers and books, but no major output in genetics or in the medical field in general. Judging by the structure of the Center, this was expected.

The Center is run by an Executive Board led by Deputy Ruler and UAE Minister of Finance, Sheikh Hamdan bin Rashad al-Maktoom. The other members are nationals with doctorate degrees from Western Universities and faculty members with specialization in micro-biology or a related fields. The vast majority of the Board members have expertise professorial in nature, with no renowned reputation as scholars, researchers, or specialists. One would have expected that given the interest of the government in purchasing the best technological products, they would do the same with the Center: hire the best lab researchers to benefit human health, as they claim, and in the process make health products less costly, as seen with the HGP.

Research and Development in Higher Education

Scientific research at Arab universities has been a low priority. The lack of funding in research development and minimal support for faculty to do research, attend conferences, and publish, and a lack of resources to carry out research, are all indications that scientific research has little strategic importance in the promotion of ICT in the sphere of education. This is best exemplified by the Internet resources available for students and faculty at learning institutions, and particularly universities. Much of the effort of Arab administrators, who are in charge of higher education, has been in acquiring a small number of desktop computers and equipping them with the Internet with limited access and a few available hours to users, hindering, as a result, the acquisition of the knowledge necessary for bridging the digital divide (Ahmed & Albuarki, 2017, pp. 28–34)

Only recently Arab governments began to incorporate a technological mode of instruction in higher learning. Skeptics, conspiracy theorists, religious moral values advocates, and concerns about immoral materials on the Internet gave the Arab governments reason to take their time in adopting the Internet (Mirza & Abdlukareem, 2011, p. 84). Forced to integrate in the digital world, some Arab countries began experimenting with the Internet as early as 1995, as in the case of the UAE, but others countries started later. In the process, Arab governments put up technical obstacles to preempt users from accessing what the governments deemed undesirable materials. Additionally, they started to put in place restrictive measures in an attempt to censor the Internet from the general public. Saudi Arabia, the UAE, Syria, and other countries created gateways and proxy servers to control the Internet while restricting its usage by creating unnecessary steps to access it. For example, in many Arab countries users have had to fill out Internet access applications at the post office and pay a processing fee.

It was with the intent to control the internet that information technology was later introduced to higher education. Even though it has been over two decades since the Internet landed in the Arab World, it is still as if it is in its infant stage. E-learning is still regarded with suspicion and many universities offer online courses as non-credit courses only. Similarly, web-design is still in its emerging stage, as many traditional universities are beginning to introduce technology to students and faculty (Al-Asmari & Khan, 2014; Weber, 2010).

Unlike the market-led research, academic research has gotten only rhetorical support by most Arab leaders. Underfunded, academic research is still in its infancy. Many barriers, including a lack of support for faculties and poor conditions for conducting research, have hindered research productivity and potentially undermined the success that market-led researchers have achieved in bridging the digital divide (Nour, 2011, p. 401). For example, the Arab World's contribution to research and development has been constant with a little below 1 percent of

TABLE 11.2. % of total GDP Spending on Research and Development, Total Number of Universities and Scientific Indexed Journals in Selected Arab Countries (1996–2011, Adapted)

Country	% of total GDP	Total Number of Universities	Total Number of Indexed Journals
Bahrain	0.2	11	2
Iraq	0.10	35	0
Jordan	0.42	29	1
Kuwait	0.09	6	3
Lebanon	0.30	24	0
Oman	0.19	11	0
Qatar	2.0	2	0
Saudi Arabia	0.05	61	6
Syria	0.20	15	0
UAE	0.15	31	12
Non-Arab Countries			
Japan	3.43	587	238
Israel	4.5	21	13
Turkey	0.7	86	54

Source: Adapted from Meo SA, Al Masri AA, Usmani AM, Memon AN, Zaidi SZ (2013) Correction: Impact of GDP, Spending on R&D, Number of Universities and Scientific Journals on Research Publications among Asian Countries. PLoS ONE8(10): 10.1371/annotation/3a739c2a-d5f2-4d6f-9e0d-890d5a54c33d. https://doi.org/10.1371/annotation/3a739c2a-d5f2-4d6f-9e0d-890d5a54c33d

the total Gross Domestic Product (GDP) for nearly three decades. In her study on science policies and higher education in Gulf and Mediterranean Arab countries, Nour (2011) noted that even though some Arab countries have intensified their efforts to increase funding for research and development (e.g., Algeria, Tunisia, and Gulf countries), they remain below the 1 percent recommended benchmark the UN set for them in 1999. (Nour, 2011, pp. 388–394).

This had a strong effect on publication productivity. Nine countries in the Arab World, including UAE, Algeria, and Saudi Arabia, produced 14,448 scientific and technical journal articles for 2000–2013. For the same period, Republic of Korea had 213,694 and Japan 318,721.

Not that the money is not there. Think tanks and Middle Eastern centers in the region and in Western countries are well-funded by several sources from the Arab World. For example, funding for the strategic centers in the UAE, Kuwait, and Saudi Arabia are unknown. The centers have produced major contributions to the literature and provided financial compensation to researchers for articles or books they publish with them. Other than including in the application the amount that the center would compensate the researcher for, there are no data available to the public showing the total amount allocated to research and development in these centers.

Furthermore, it is known among Westerners that the Gulf rulers are generous donors to education and scholarships. During the late King Abdullah bin Abdulaziz"s tenure (1924–2015), Saudi Arabia was the fourth largest donor to the scholarship fund for Saudi Arabia's international students in the U.S. Similarly, the Qatari Foundation, under the leadership of the Emir Sheikh Tamim bin Hamad Al Thani, has provided full scholarships to thousands of university students interested in learning and teaching Arabic as a foreign language in U.S. schools. The same applies to the support of many Arab leaders for the established Middle Eastern centers in the U.S., Canada, and Europe. Without such support these centers would not survive under the constraints of budgetary cuts and economic crises that these countries have gone through in recent decades (Nour, 2011, p. 391).

Certainly, there is value to investments in think tanks and Middle Eastern centers outside the Arab region. Nevertheless, up to now, think tanks and strategic centers in the Arab World remain isolated from the universities and restricted from the public at large. These institutions require invitations to attend their public gatherings, and their discussions of most strategic issues are held in secret and away from the public eye or the media. Similarly, conferences and forums organized by Middle Eastern centers in the Arab World are scripted for political correctness. For example, reports issued by the center for International Institute of Strategic Studies in Bahrain do not address the regional issues or the political turmoil that may involve or have implications on their host country, Bahrain. Several Western scholars have raised questions about scholars' ethical standards and responsibility to acknowledge when they censor themselves out of fear of repercussion to themselves or the center they work for. Davidson (2013), Coates-

Ulrichsen (2013) and Lindsey (2012) have discussed extensively the problems scholars encounter when they express views not necessarily in line with what the governments of the Arab states of the Gulf region consider "legitimate criticism" (Coates-Ulrichsen, 2013). In February 2013, a conference at the American University of Sharjah (AUS), organized by the London School of Economic and AUS, was cancelled after it was determined that AUS would not allow Kristian Coates-Ulrichsen to present her paper on the uprising in Bahrain, at the conference. Later, Coates-Ulrichsen was denied entry to the UAE for the same reason. According to Coates-Ulrichsen, in an "age of austerity budget slashing" and cutting in the West, universities and academics may find themselves expose to "new pressures and vulnerabilities." (Coates-Urichsen, 2013).

To a degree, a similar concern is raised towards offshore educational institutions established recently in the Gulf countries. The Gulf States are leading the region, benefitting from the establishment of foreign affiliate universities and foreign universities' branch campuses. These offshore institutions are administered by local authorities, but their curricula are the same as those in the home country, and they offer degrees recognized in the affiliate university. (Miller-Idriss & Hanauer, 2011, p. 185) These phenomena have expanded transnational educational opportunities for the Gulf States (Miller-Idriss & Hanauer, 2011, p. 185) and contributed to access to e-learning modalities and, overall, increase of Internet penetration in these states. As of now, the total number offshore institutions in the Arab World is 57, with 50 of them located in the Gulf States.

Funded by the host country and adhering to their home country's demanding research agendas (as in the motto "publish or perish"), offshore institutions have had a strong impact on research productivity. Looking at the impact on publications, the UNESCO 2030 Report shows a significant increase in the total number of published articles in which more than one author from the Arab World contributed to the final product. The data reveal a pattern showing transnational collaboration as a reflection of historical relations and political agendas that exist between the Arab collaborators and those of the outside. In this case, the U.S. and France have the highest total number of published articles from 1996–2011 at 19,193 and 15,831 respectively, followed by Saudi Arabia with 10,110 for the same period. What also emerges from these data is a reproduction of potentially an educational system dependent on old colonial and new strategic allies. This is more pronounced in Northern African countries. For example, Morocco, Algeria, and Tunisia are strong partners of France. Over 90 percent of all collaborations are with French authors. In the language of dependency theory, this collaboration could perpetuate an undesirable educational system that is best suited to the benefit of France than those of the Northern Africa countries. In this sense, collaboration would become a means to foster a dependent relationship that would ultimately make the advancement of the Northern African countries on the technological stage more reliant on France, preempting the Northern African educational system from developing its own transitions. In the end, this could be a

TABLE 11.3.   International Collaboration

Host Country: USA			
Jordan	1153	Egypt	4725
Algeria	383	Tunisia	544
Iraq	279	Kuwait	566
Saudi Arabia	5794	Morocco	833
Bahrain	89	Lebanon	1307
Mauritania	18	Sudan	185
Oman	333	Palestine	35
Syria	170	Qatar	1168
Yemen	106	UAE	1505
	Total 19,193		
Host Country France			
Algeria	4883	Lebanon	1277
Mauritania	62	Morocco	3465
Syria	193	Tunisia	5951
	Total 15,831		
Host Country: Saudi Arabia			
Bahrain	137	Algeria	524
Jordan	490	Tunisia	600
Kuwait	185	Egypt	7803
Yemen	158	Sudan	213
	Total 10,110		
Host Country: Egypt			
Palestine	50	Bahrain	101
UAE	370	Saudi Arabia	7803
Kuwait	332	Libya	166
Yemen	183		
	Total 9,005		

Source: Adapted from UNSCO 2030 Report

cause for furthering the digital divide within each of the Northern African states even though internationally the digital divide appears narrowing in relation to the outside world.

There are serious concerns among Arab scholars and experts in higher education about these offshore institutions. Bashshour (2010) and Akkary (2014) question the motives of the institutions and see them as tools to promote technological practices devoted to the promotion of free trade and consumerism not necessarily congruent with the educational mission of higher learning expected to come with them. Accordingly, there is a consensus that international agencies and Western universities are driven by the home countries' economic incentives, using the Gulf countries as a source for generating revenue from enrollment of foreign students and donors in the Gulf leaders. Bashshur characterizes this phenomenon as "an educational revolutionary adventure." He notes how it traps the region in a full cycle of dependency, by transplanting American ideas into the Arab context uncritically, and through the perception that the prepackaged foreign university curriculum and educational programs are good enough to resolve the Arab World's educational shortcomings because they are simply from the advanced countries (Bashshur, 2010, p. 260).

What is more worrisome for Arab scholars and experts is the marginalization of domestic faculty that takes place with the creation of these offshore entities. Even though these institutions are administrated and managed by the host country, foreign educational consultants are in abundance, solicited for their expertise in the development and execution of the offshore —"modern"—educational model. The consultants are expected to teach technical—"Western"—courses and advise on technical matters. Western expatriates who worked in the UAE higher education system report that they were never asked to "consider religiously or culturally sensitive areas of the curriculum" (Findlow, 2005, p. 290). At the same time, the local faculties are kept outside the process, unable to contribute to the cultural adaptation of the imported educational model.

This has led Arab educational reformers to wonder whether there is a real educational benefit to transplanting Western-based universities to the Arab World. To them the question of benefit becomes a matter of survival. When faced with this strategic question, Bashshur simply replies pondering: "How [can one] benefit from the West without crushing under its weight and losing one's soul and heart in the process?" (Bashshur, 2010, p. 269).

CONCLUSION

Considering the data gathered for this study, there is enough evidence that the Arab World is transitioning to the Digital Age at an uneven speed. The high-income countries of the GCC have done well in terms of access and usage of technology, but the rest of the region remains far behind. The evidence also reveals that market and profit are driving the incorporation of ICT and that the UAE and Qatar have narrowed the digital divide in their favor. But, upon closer examination, the data also show that such a focus may result in undermining the Arab state to sustain this growth and potentially become trapped in a dependent relationship that would undermine its capacity to narrow the digital divide. Additionally, the low priority assigned to research and development in academia as a reliable source for preparing the next generation to lead the process of development is proving to be a reason for the inability of the Arab World to bridge the gap of the digital divide. Such a lack of emphasis on research will likely result in growing the gap even further in the future.

REFERENCES

Akkary, R. K. (2014). Facing the challenges of educational reforms in the Arab world. *Journal of Educational Change, 15*(2), 179–202.

Al-Asmari, M. A., & Khan, M.S. (2014). E-learning in Saudi Arabia: Past, present and future. *Near and Middle Eastern Journal of Research in Education.* Retrieved from, http://dx.doi.org/10.5339/nmejre

Al-Emran, H., Elsherif, M., & Kaled S, (2016). Investigating attitudes towards the use of mobile learning in higher education. *Computers in Human Behavior, 56*(2), 93–102.

Algadheeb, N. A., &.Almeqren, M.A. (2014). Obstacles to scientific research in light of a number of variables. *Journal of International Education Research, 10*(2), 101–110.

Alrawabdeh, W. (2009). Internet and the Arab world: Understanding the key issues and overcoming the barriers. *The International Journal of Information Technology, 6*(1), 27–33

Bashshur, M. (2010). Observations from the edge of the deluge: Are we going too fast in our educational transformation in the Arab Gulf? In O. Abi-Mershad (Ed.), *Trajectories of education in the Arab World: Legacies and challenges* (pp. 247–272). London, UK, and New York, NY: Routledge in association with the Center for Contemporary Arab Studies, Georgetown University.

Benzuidenhout, L. M, Leonelli,S., Klly,A.H., & Rappert, B. (2017). Beyond the digital divide: Towards a situated approach to open data. *Science and Public Policy, 44*(4), 464–475.

Brookings Institution. (2008). *A new millennium: Human development in the Arab world.* Washington D.C.

Coates-Ulrichsen, K. (2013). Arab upheavals trigger new sensitivities in Gulf funding for western universities. *Scholars at Risk Network.* Online edition. July 1. Retrieved from: https://www.scholarsatrisk.org/resources/arab-upheavals-trigger-new-sensitivities-in-gulf-funding-for-western-universities/

Davidson, C. (2013). *After the Sheikhs. The coming collapse of Gulf monarchies.* Oxford, UK: Oxford University Press

Dutta, S., & Coury, M. E. (2003). *ICT challenges for the Arab World.* Retrieved from https://pdfs.semanticscholar.org/9672/b63ae09331827d4678a8acfb06cd55bc643d.pdf

Findlow, S. (2005). International networking in the United Arab Emirates higher education system: Global-local tensions. *Compare, 35*(3), 285–302.

Gelvanovska, N., Rogy, M., & Rossotto, C. M. (2014). *Broadband networks in the Middle East and North Africa. Accelerating high-speed internet access.* Washington DC: the World Bank.

Heusel, J., & Richards, N. (2017/2018). Why precision medicine has to rely on big data. *Wired Magazine,* 51.

Kaba, A., & Said S. (2013). Bridging the digital divide through ICT: A comparative study of countries of the Gulf Cooperation Council, ASEAN and other Arab countries. *Information Development, 30*(4), 358–365.

Klischewiski, R. (2014). When virtual reality meets real politik: Social media shaping the Arab government-citizen relationship. *Government Information Quarterly, 31,* 358–364.

Lindsey, U. (2012). *NYU-Abu Dhabi behaves like careful guest in foreign land.* Retrieved from, https://www.chronicle.com/article/a-careful-guest-in-a-foreign/132075

Loch, K. D., Detmar W. S., & Sherif, K. (2003). Diffusing the internet in the Arab world: The role of social norms and technological culturation. *IEEE Transactions on Engineering Management, 50*(1), 45–63

Martin, H. (2011). The end justifies the definition: The manifold outlooks on the digital divide and their practical usefulness for policy-making. *Telecommunication Policy, 35,* 715–736.

McGarty, C., Thomas, E.F., Lala, G., Smith. L. G., & Bliuc, A. M. (2013). New technologies, new identities, and the growth of mass opposition in the Arab Spring. *Political Psychology, 35*(6), 725–740

Miller-Idriss, C., & Hanauer. E. (2011). Transnational higher education: Offshore campuses in the Middle East. *Comparative Education, 47*(2), 181–207

Murphy, E. C. (2009). Theorizing ICTs in the Arab world: Informational capitalism and the public sphere. *International Studies Quarterly, 53*(4), 1131–1153.

Nour, S. S. O. M. (2011). National, regional and global perspectives of higher education and science policies in the Arab region. *Minerva, 49,* 387–423

Pearce, K. E., & Ronald E. Rice, R. E. (2013). Digital divides from access to activities: Comparing mobile and personal computer internet users. *Journal of Communication, 63,* 721–743.

Shamaileh, A. (2016). Am I equal? Internet access and perceptions of female political leadership ability in the Arab world. *Journal of Information Technology & Politics* [AU: volume, number, page range]. Retrieved from, http://www.ammarshamaileh.com/Shamaileh-Am%20I%20Equal.pdf

UNDP (2011). *Arab Knowledge Report 2010/2011. Preparing Future Generations for the Knowledge Society.* United Arab Emirates, Dubai.

UNESCO Publishing, (2015). *UNESCO Science Report, Towards 2030.* Paris, France: Author.

Weber, A. S. (2010). *Web-based learning in Qatar and the GCC States.* Qatar: Center for International and Regional Studies. Georgetown University School of Foreign Service. Occasional Papers.

CHAPTER 12

ASSISTIVE TECHNOLOGY FOR STUDENTS WITH DISABILITIES

An International and Intersectional Approach

Saili S. Kulkarni, Jessica Parmar, Ann Selmi, and Avi Mendelson

Based on a review of almost two decades of research, augmentative and alternative communication devices tend to be promoted and discussed using a mostly dominant framing of race, culture, language and context (Kulkarni & Parmar, 2017). These communication devices are part of a broader set of assistive technologies (AT) supports provided to students with disabilities primarily in school settings. While devices can be meaningful for the participation, socialization, and employment of students with disabilities, particularly those with difficulties with verbal speech and mobility, few studies have highlighted important perspectives of culturally and linguistically diverse student users, their families and community members. In this chapter, we call for a framework of intersectionality (Crenshaw, 1991) that embraces race, culture and language when supporting students with disabilities who utilize assistive technology devices. We begin by introducing AT use and research internationally, addressing some of the underlying issues around participant demographics (mainly that intersectional markers are often not included), providing 1–2 case examples of students with disabilities at the intersections of race and language who utilize AT, and concluding by providing a framework for future research on culturally and linguistically diverse AT research.

Crossing the Bridge of the Digital Divide: A Walk with Global Leaders, pages 197–208.
197

Assistive technology (AT) includes a set of supports and devices that accommodate a variety of learning needs and differences. These devices and supports traditionally include a universally designed learning environment (UDL) and have been shown to transform student success in the schools (Harper, Kurtzworth-Keen, & Marable, 2017). Further, AT also has the potential to provide accessibility to individuals with identified disabilities. Currently, however, AT research, especially as it includes "cutting edge" technology, has been limited. Particularly, there are very few empirical studies highlighting new possibilities for technology exist (Wehmeyer, Palmer, Smith, Davies, & Stock, 2008).

AT devices can enhance or maintain an individual's existing skills or compensate for nonexistent or absent skills (Edyburn, 2004). AT can include several high and low-tech options that support students with disabilities in school settings. Low-tech options include items that are considered easy to use, that do not require extensive training, and that are low-cost / easily available (Edyburn, 2004). These items are traditionally marketed widely and include very simple gadgets. Low tech examples include pencil grips, individual Whiteboards, binders for slanted/raised writing, raised lines on paper, Velcro enhanced books, highlighter tape, laminated highlighting, rubber stamps, visual timer, and so forth. High tech AT options include options that are usually costly, and often require evaluation for fit and training for users and family members. These can include augmentative and alternative devices (AAC), software, iPads, computers, and other devices that individuals may require supporting learning.

Research related to AAC devices has shown that these devices are often approached from dominant, Westernized contextual frameworks of understanding and implementation techniques, particularly when they are used in conjunction with schooling. Globally, AAC devices are still researched from dominant European-American perspectives, training is unavailable in languages other than English, and speech-to-text options on such devices is produced in English (Kulkarni & Parmar, 2017). The challenge of supporting the implementation of AAC devices and other high and low tech AT options in school settings, therefore, is that they are often not tailored to most students. Schools in the United States and United Kingdom, for example, have seen a surge of non-dominant groups of students attending schools in the past ten years (Warner & Palfreyman, 2001).

In the United Kingdom, there is no national strategy for assistive technology. While equipment may be funded by the Local Education Fund (LEF) for students in public schools, there is no standardized set of practices that govern assistive technology. Additionally, in higher education contexts, the Disabled Students Allowance provides the purchasing of equipment deemed as AT (McKnight, 2016). However, similar to the United States, obtaining equipment can be a lengthy process and supports do not necessarily transfer from one environmental context to another (McKnight).

While children with disabilities in the digital age can use technologies to readily and efficiently connect with formerly denied areas of cultural exchange, these technologies can also present barriers and inaccessibility issues similarly faced by adults with disabilities (Alper & Goggin, 2017). This is not just an area of concern for basic accessibility. Rather, as Alper and Goggin (2017) explain, it also affects "the ecosystems, economies, platforms, formats, and the very nature of communication are often entwined in the social shaping of disability and digital media" (p. 730). While advancements in technology have been made for educational contexts, such advancements are often not available to students with disabilities mainly because of issues such as cost, lack disabled individuals' involvement in the design process, attitudes of the community, and issues with support providers who are not trained to use devices (Darcy, Green, & Maxwell, 2016). All of these issues are amplified, however, for individuals who identify as culturally and linguistically diverse and are at the intersection of race, culture, language and disability.

Intersectionality

Traditionally, issues tied to assistive technology have maintained dominant, Euro-American ideals of accessibility, perception and implementation. In this chapter, we utilize an intersectional framework to further understanding of how such devices may be accessed, perceived and implemented among culturally and linguistically diverse groups of individuals. We specifically draw on Crenshaw (1991) who explains that "The problem with identity politics is not that it fails to transcend difference, as some critics charge, but rather the opposite-that it frequently conflates or ignores intragroup differences" (p. 1242). We claim that such differences serve an important role in intragroup differences among individuals with disabilities as they utilize AT.

Although Crenshaw's (1991) original ideas about intersectional identities focused on the interactions between gender and race, particularly through structural and physical violence against women of color, Erevelles and Minear (2010) move us toward an intersectional framework that also includes disability as identity. In particular, Erevelles and Minear (2010) point out the dangers of the missing identity of disability from conversations including critical, social justice frameworks, particularly as they relate to educational contexts. It is often in educational contexts where students with disabilities at the intersections of race, culture and language are often segregated from 'typical' peers and often experience detrimental outcomes that contribute to underemployment and the school to prison pipeline (Artiles, 2013; Erevelles & Minear, 2010).

Therefore, we see intersectionality as an important lens for all topics involving the education of students with disabilities, particularly those students of color who are often doubly marginalized as disabled and raced (Connor, Ferri, & Annamma, 2015). Such students' access to meaningful AT devices that support and enhance their learning opportunities are also understudied in the research. In the subsequent section, we provide some examples both of students in school settings and of their families who have experienced such marginalization in the United States.

MARTIN'S LABEL MAKER

Martin is a seven-year old African American boy with cerebral palsy. He is in the 2nd grade at a public elementary school in Northern California. Martin was identified for special education services at age 4, through the regional center. During kindergarten, Martin attended a segregated class with mostly African American and Latino/a students, which focused on learning both appropriate school behaviors and functional life skills. At his annual Individualized Education Program (IEP) review meeting, teachers and service providers including speech/language, physical therapy, occupational therapy and special education teachers all believed that Martin was "misplaced" in a segregated classroom and encouraged his parents to opt for an inclusive elementary program.

Martin was bused for an hour each morning to attend a public elementary school that provided an inclusive elementary program experience. During his first few days in the 1st grade, Martin was made aware that he would have both a general education classroom teacher that he shared with about 20 other students and also an itinerant special education teacher who would provide him with additional supports in class. He would maintain his other services including speech/language, occupational, and physical therapy, often leaving his classroom up to three times each week for services.

Martin was a happy, positive child. In spite of his use of a wheelchair to move about the school, Martin easily made friends in his first grade class with some gentle facilitation by his general and special education teachers. Academically, Martin was comfortable in his classroom. He readily answered questions, and made efforts to complete his homework assignments. However, school was difficult for Martin during handwriting time. While his peers sloppily added letters to large lined paper, he anxiously held his pencil as his hands shakily approached the page. His general education teacher often remarked that Martin became distracted during handwriting time and would try to avoid completing his activities.

The special education and general education teacher called a meeting with the occupational therapist and father to discuss whether it might be possible for Martin to receive some additional handwriting supports for his schoolwork. The occupational therapist suggested a label maker to assist Martin with some of his more difficult writing tasks in class. To receive the label maker, Martin had to undergo several more formal assessments, first by the occupational therapist, and then by the AT district coordinator who had to see if it was something the district could and would purchase for Martin to use.

Martin's father was unfamiliar with this labored process of obtaining the label maker and was unsure about how to proceed initially. This uncertainty led to some negative perceptions of Martin's father by the special education staff and occupational therapist. They were inclined to say that Martin's father "didn't care" about his education. In reality, Martin's father, a single dad of three kids, worked late night shifts to provide for his family. Martin was often left with his grandmother and siblings while his father worked. Despite his hectic schedule,

Martin's father continuously provided snacks for Martin's classmates, cupcakes for Martin to share on his birthday, and chaperoned the class trip to the zoo at the end of the school year.

With the help of Martin's special education teacher, who advocated for the label maker alongside the occupational therapist, the AT specialist was able to communicate with both Martin and his father about the process of having the school purchase Martin's label maker. Even after all of the assessments were completed, it took close to a year for Martin to receive his label maker with because of the mounds mountains of overdue paperwork to complete for the AT coordinator (the only one for a large district).

Eventually, however, Martin received his label maker and was able to use this it to combat some of the frustration that he felt during handwriting time in class. The label maker allowed Martin to type out longer phrases at a time. With the help of an aide, he was able to paste these sections on his homework and complete his work more efficiently.

AT at the Intersection of Disability and Race

Martin's story illustrates some of the frustrations faced by students and families when attempting to access AT devices in the U.S. school system. In particular, negative attitudes towards uninformed families when navigating the AT process, the difficulties with obtaining devices given the time length for individualized AT assessments, and the backlog of paperwork experienced by the AT specialist, all contributed to delays in Martin's access to meaningful participation in his inclusive classroom during handwriting time. Martin's story, therefore, highlights three key considerations for an intersectional approach to AT: (1) meaningful family participation and relationships during the AT assessment process (McCord & Soto, 2004), (2) streamlined assessment procedures for student access, (3) and, culturally responsive instruction for service providers, who tended to, in Martin's case, make assumptions about the involvement of his father (Kulkarni & Parmar, 2017).

JUAN'S HIGH TECH SPEECH OUTPUT

Juan is an 11-year-old, 5th grade student grader with cerebral palsy attending an elementary school in Northern California. Juan had been a part of inclusive education programs since kindergarten. Juan was a social butterfly in classes, loved the attention of classmates and would bring candy to class in order to trade. Juan's grandmother and grandfather maintained custody of Juan, as his mother had gone to jail and his father was not involved in his life. His grandmother and grandfather spoke fluent Spanish, but very little English. They often had a difficult time communicating with school professionals and at his IEP meetings, where the majority of professionals only knew English. Juan's 4th grade teacher was a male from Chile and was very enthusiastic about Juan's presence in the classroom. He ensured that Juan was present and actively engaged in the classroom, and he was able to communicate fluently with Juan's grandparents about his yearly progress.

Unfortunately, when Juan entered the 5th grade, his speech began to decline as the muscle tone in his mouth increased slightly. His physical therapist and speech therapist both noted this change. This meant that while Juan was able to have meaningful interactions through his inclusive setting and complete his coursework through the assistance of a scribe, his ability to communicate verbally with adults and peers was exceedingly difficult. His special education teacher decided to talk to the AT coordinator about getting Juan assessed for a high-tech communication device. With the support of the speech therapist and special education teacher, Juan was assessed and provided a Tobii Dynavox T7[1] ®. The major function of this device was to generate complex speech for Juan. Juan worked with the speech therapist, special education teacher, and AT coordinator to learn how to use the device.

In the classroom, however, the general education teacher had no training in how to facilitate any questions that arose as a result of Juan's device. Students' natural curiosities meant that Juan was constantly having peers leaning over his wheelchair to see his new device. Juan initially liked the attention, but eventually became overwhelmed by the unnatural repeated invasion of his personal space. While this new device was supposed to provide access, communication and social opportunities with his peers, most saw the device as a novel toy and only approached him for this reason.

In addition to the issues with facilitating meaningful conversations in the classroom about Juan's device, there was the added challenge that his grandparents were not given the same time, consideration and support to learn how to use the device at home. Therefore, when Juan first took home his device, it would remain in his room unused and sometimes he would forget to bring it to school. The district coordinator and speech therapist took this as a sign that the grandparents were negligent and instructed that the device should only remain at school.

Juan's T7 device did not include any language translation options. Juan's grandparents were not given any training on the device before accepting it as a part of Juan's daily routines. These inequities tend to lead to misuse or abandonment (Kulkarni & Parmar, 2017). Juan's grandparents were not even given the opportunity to learn about his communication device before a gatekeeping mechanism was used to prevent them from having further interaction with it. Unfortunately, this is an all too common occurrence in the school system when it comes to the relationships between school and culturally and linguistically diverse families (Artiles & Trent, 2013). The presumed incompetence of Juan's family illustrates some of the barriers that families face in obtaining necessary resources for their children with disabilities.

---

[1] A symbol adapted speech output software.

AT at the Intersection of Language and Disability

Juan's story, therefore, highlights additional considerations for students who require high-tech AT options to be successful in school: (1) the importance of meaningful instruction for peer partners, all teachers, and service providers (McCord & Soto, 2004); (2) opportunities for family members to be presumed competent and given opportunities to learn about AT supports in order to implement these in the home; and (3) the need for further research and communication device options in languages other than English (Kulkarni & Parmar, 2017). While in both cases, AT coordinators and assessors were seen to be overworked, overburdened and in need of support, the lack of a multidisciplinary team approach to Juan's T7 device meant that his grandparents were unable to support him in maintaining communication in the at home. Additionally, given his grandparents' first language was not accommodated in the context of the choosing and implementation of the device, it becomes more important that an intersectional AT approach accommodate intersections of disability and language.

JAMES' AAC DEVICE

James is a 13-year old, 8th grade student with Autism attending a middle school in Southern California. In addition, James is nonverbal, and was provided with an AAC device in the 2nd grade to access curriculum, and to access basic communication/socialization needs. James' immediate family consists of 1 younger brother, both grandparents, two aunts, a caretaker, and his mother, all of which live with him at home. James was diagnosed with Autism when he was 3 years old, and has attended schools in in a local school district since pre-school. He has been a part of Special Day Classes, which only include a wide variety of students with disabilities. Since James was a baby, he has had his stay-at-home caretaker to assist with drop-off at school, meals at home, and pick-up at school. The family requires this additional assistance, as the dad is not present in James' life, and James' mother has always had to work to support the family. James' mother is the only person who works in a household consisting of 8 people. His family speaks Spanish, and very little English. The family has never worked with a teacher or service provider that speaks Spanish, and so communication with schools does not occur as often as it should.

James was known as one of the few students who had an "easy-going" family. In addition, James was often a well-behaved, compliant student. However, once James reached 5th grade, James began engaging in problem-behavior in the classroom. This was difficult for the classroom teacher to support, as staff members were unable to determine the function of his behavior. There would be times where James was eating a snack—a preferred activity—and suddenly he would throw his food to the ground and run across the playground. He would cry during this time, or physically lash out at those who were within proximity and attempted to calm him down. The only strategies he responded to were breaks on a cushion, where he was able to lay down and have space away from others. Eventually, James' tantrum behavior became excessive in intensity, as well as in length, and an emergency IEP was held. An aide named Angie was assigned to him for behavior support.

When Angie was assigned to James, her initial observation was that James had no way to indicate either his feelings, physical state, or if he needed something. Rather than respond to his problem behavior, Angie wanted to prevent it. She began to assist in developing strategies to assist James in expressing his basic needs, such as a simple "break card" that James would hand to staff members to indicate he needed a break instead of having tantrum behaviors. However, she also noticed James had an iPad that he was using to watch YouTube. When Angie asked what his iPad was for, the speech pathologist explained to her that this was an Augmentative and Alternative Communication (AAC) device that was issued several years ago for James to communicate. He had an application on it called Proloquo2Go, with several basic words within it. Some of these expressions included "break," "snack," and "home," with images assigned to them. Angie decided she wanted to have him try and use the AAC device with her, as another support for requesting a break.

When Angie was with James, she also began to notice that James' was working on academic and social areas that were not challenging or near his instructional level. James was practicing receptively identifying vocabulary, because his IEP stated that James was unable to read or communicate functionally. As Angie continued to ask other staff members questions in order to better support him, Angie was told that there was no use in trying to get James to communicate with his iPad because he would use it as a toy and echo the basic words from the AAC back at others instead of using it functionally. In addition to this, she was told that he did not speak English or Spanish, but that he liked to "babble." Many people that worked with James previously felt that James was unable to progress because of the language barrier, and insisted that living in a Spanish-speaking household was "confusing" for James.

Angie was in a university-based special education teacher credential program for moderate to severe disabilities at the time, and began to see that James was a student not being pushed to his full potential and was experiencing a disservice due to a his non-dominant cultural and linguistic background. Angie decided that she wanted to really challenge James across all academic, social, and functional areas and began to advocate for him alongside the classroom teacher.

In Angie's credential program, she was taught that helping a student develop their own voice was key to their overall education, social, and emotional well-being. She also knew that current research shows that students with exposure to more than one language tend to have better learning outcomes. Angie decided to test the theory that James used his AAC device as a toy, and spent several days pretending she did not understand him unless he was using his AAC device to communicate. Within one week, James had gone from repeatedly pressing random icons on his device, to using it to request his wants and needs, such as favorite foods, the bathroom, and breaks during instruction. After two weeks of consistent use in getting his needs met, James started using verbal expressive language to request things as well. Angie was so enthralled with this progress, that she began to add curriculum-based vocabulary to James' device. As he continued to pick up this language as well, she tried one more test and took the pictures out of his device to see if he could read the words. Within another several months, James was reading CVC sight words and the classroom teacher had started to introduce a sight-word reading program for James. However, the biggest success from James' use of the AAC device was when Angie added a section for feelings. She added pictures and words for "mad," "sad," "happy," and "sleepy." Before James felt the need to have a tantrum behavior, he started to take out his AAC device and press the icon for "mad." This prevented most of his behaviors, and Angie was able to give James breaks. One day, he pressed the icon for "mad" repeatedly and found the word for "mouth" in his device, which had been added on another day when they were learning body parts. This information combined helped everyone realize that James was mad because he was having pain in his teeth. When his mother took him to the dentist for this, James ended up needing to have several cavities filled. James' access to the AAC device finally gave everyone a direct reason for why James had been acting out.

James went from "babbling" and using his device as a toy, to forming expressive language, getting his needs met, and using his device to answer questions in class. However, the barrier that still remained was the language barrier between the home and school. Though Angie knew some Spanish, she did not know what was most familiar to James at home, but added some simple phrases to his device. She wished the boundaries of her role as a support staff allowed her to work with the family in adding language supports to the device. Though James's journey is a success, his story situation also emphasizes the need for a more profound understanding of language and culture, and how these areas intersect amongst collaboration with family and professionals.

AT at the Intersection of Language and Disability

As evident in James' story there are extensive challenges of working with multicultural and multilingual students. Many professionals view bilingualism as a challenge to overcome, rather than as a value. When a student from a non-dominant culture has a disability, is bilingual, and non-verbal, these factors can put the student at a disadvantage for receiving instruction that is meaningful. Not only did the staff at James' school view his Spanish-speaking background as "confusing" for him, but they additionally continued to share with Angie that James was a lost cause when it came to communicating functionally. This attitude stems from lack of training and support in the area of AAC devices for classroom teachers as well as support staff, as it seems that nobody in the classroom knew how to access and appropriately assist James in use of the device. James' also came from a household where the cultural norm is for the family to fully respect the educational decisions of the school, and to be uninvolved. Had James come from a family that was given more support and training in use of AAC, more access to opportunities to collaborate with the classroom teacher with translator support, and encouragement to advocate for their son, James may have received intervention at a much earlier point. When both the family and professionals had no idea how to use the device, the result was to view it as a toy and abandon communication opportunities. James' story is a call for attention to the intersection of culture, language, and race, as well as the importance of respectful family-professional relationships.

DISCUSSION AND CONCLUSION

This chapter provided an overview of some of the issues surrounding AT in the educational systems both within the United States context and a more general discussion of international issues. We argued that among these issues related to access and implementation of AT devices within school and other environmental contexts, an equally important issue is the need to understand how disability, race, culture, and language intersections influence AT adoption and implementation. Just as AAC devices tend to be positioned through Westernized perspectives (Kulkarni & Parmar, 2017), AT devices need to be understood through intersectional lenses. Using the American-based cases of Martin, Juan, and James, we illustrated how dominant frameworks of understanding students with AT needs and their stories are limited in scope. Instead, we posture that AT devices need to be thoughtful in highlighting the complexities and intersections among disability, race, culture and language.

Recommendations

We recommend several practices that capitalize on the framework of intersectionality introduced by Crenshaw (1991). First, as learned from Martin's story, meaningful family participation and relationships during the AT assessment process means that discussions with families are *valued* as equal to or higher than those of professionals. This radical notion capitalizes on the fact that *all* families value education and support for their children in the school system and bring 'funds of knowledge' into the school setting. As Moll, Amanti, Neff, and Gonzalez (2001) explain, funds of knowledge refer to "historically accumulated and culturally developed bodies of knowledge and skills essential for household or individual functioning and well-being" (p. 133). We believe that in all of the represented cases, that professionals should appreciate these sets of knowledge and experiences brought in by families of children who are being evaluated for or utilize AT devices.

Second, we strongly encourage the recruitment and retention of special education teachers who are committed to radical forms of social justice. These teachers have a heightened awareness of the various differences that students at the intersections of disability, race, culture, and language bring into the classroom and are able to advocate for AT devices and supports that integrate these intersectional identities. In addition to this, we recommend the recruitment and retention of linguistically diverse teachers and service providers who can support linguistically diverse families navigating the AT assessment, access and implementation process. It is also imperative that research begin to account for AT devices that utilize languages other than English and visual representations of culturally and linguistically diverse individuals. Most importantly, teachers and professionals must bridge relationships with students, family members, and multidisciplinary team members to encourage the use of AT devices that include meaningful participation in school and the community.

As was suggested in the literature and our case studies, AT device assessments and policies should be streamlined to allow students to access them because a digital divide exists among those who have access to supports and training for such devices. Because there are limited policies for AT devices in schools internationally (McKnight), students often run the risk of not having full access to instruction due to backlogs. The hiring of culturally responsive professionals to assess and support AT implementation is an incredible need in schools, especially those that are under resourced around the world! Thus we believe that these recommendations will assist in the construction of a more intersectional approach and support a broader population of individuals with disabilities globally who utilize AT devices, their families, school professionals and decision makers.

REFERENCES

Alper, M., & Goggin, G. (2017). Digital technology and rights in the lives of children with disabilities. *New Media & Society, 19*(5), 726–740.

Artiles, A. J. (2013). Untangling the racialization of disabilities. *Du Bois Review: Social Science Research on Race, 10*(2), 329–347.

Artiles, A. J., & Trent, S. O. (2013). Culturally/linguistically diverse students, representation of. *Encyclopedia of special education: A reference for the education of children, adolescents, and adults with disabilities and other exceptional individuals.*

Connor, D. Ferri, B. A., & Annamma, S.A. (2015) *DisCrit: Disability studies and critical race theory in education.* New York, NY: Teachers College Press.

Crenshaw, K. (1991). Mapping the margins: Intersectionality, identity politics, and violence against women of color. *Stanford Law Review, 43*(6), 1241–1299.

Darcy, S., Maxwell, H., & Green, J. (2016). Disability citizenship and independence through mobile technology? A study exploring adoption and use of a mobile technology platform. *Disability & Society, 31*(4), 497–519.

Edyburn, D. L. (2004). Rethinking assistive technology. *Special Education Technology Practice, 5*(4), 16–23.

Erevelles, N., Minear, A. (2010). Unspeakable offenses: Untangling race and disability in discourses of intersectionality. *Journal of Literary & Cultural Disability Studies, 4*(2), 127–145.

Harper, K. A., Kurtzworth-Keen, K., & Marable, M. A. (2017). Assistive technology for students with learning disabilities: A glimpse of the livescribe pen and its impact on homework completion. *Education and Information Technologies*, *22*(5), 2471–2483.

Kulkarni, S. S., & Parmar, J. (2017). Culturally and linguistically diverse student and family perspectives of AAC. *Augmentative and Alternative Communication*, *33*(3), 170–180.

McCord, M. S., & Soto, G. (2004). Perceptions of AAC: An ethnographic investigation of Mexican-American families. *Augmentative and Alternative Communication*, *20*(4), 209–227.

McKnight, L. (2016). The case for mobile devices as assistive learning technologies: A literature review. *International Journal of Mobile Human Computer Interaction (IJMHCI)*, *6*(3), 1–15.

Warner, D., & Palfreyman, D. (2001). *The state of UK higher education: Managing change and diversity*. Philadelphia, PA: Open University Press.

Wehmeyer, M. L., Palmer, S. B., Smith, S. J., Davies, D. K., & Stock, S. (2008). The efficacy of technology use by people with intellectual disability: A single-subject design meta-analysis. *Journal of Special Education Technology*, *23*(3), 21–30.

CHAPTER 13

ONLINE RESOURCE COURSES TO ENHANCE EDUCATION ABROAD LEARNING

The Digital & Enhanced International Learning Divide

Gary M. Rhodes and Rosalind Latiner Raby

The mobility of university students from home countries to countries outside their home university campus continues to grow. Over 1 million international students study at US colleges and universities and over 360,000 US students at US colleges and universities study outside of the US for full degrees or as a part of their US degree program (2017 IIE Open Doors, published in 2016). Research has shown that along with international learning, student mobility can serve to enhance retention and success as well as career development for university students. This chapter will provide information about some of digital, e-learning, and social media tools and platforms that have recently been developed to support pre-departure, while-abroad, and re-entry learning for students who take part in international student mobility programs. Particular platforms will be highlighted, including IStudent101.com for undergraduate and graduate students going from any country to any other country, Ustudy.us for international students studying in the US, and GlobalScholar.us for US students studying in countries outside the U.S.

Crossing the Bridge of the Digital Divide: A Walk with Global Leaders, pages 209–225.

209

INTRODUCTION

In the last half-century, massification policies have broadened mass university access. At the same time, near universal secondary school graduation has increased competition for limited spaces and created a context in which the elite university is unable to absorb the growing demand for admission from non-traditional students. The non-traditional student is defined by one or more of the following characteristics, low SES levels, students of color, being first-generation, working full-time, being academically under-prepared, and being an adult (Ross-Gordon, 2011). While access to higher education remains important, just attending a college or university is not sufficient. Higher education leaders and researchers are increasingly focusing on what students learn while they are in higher education, whether they complete their education, whether they do so in a timely manner, and how well they succeed following the completion of their studies. For California community colleges and California State Universities that provide a lower-cost option for non-traditional students to attend higher education, emphasis is being placed on initiatives that lead to retention and success.

In both these institutional types that support massification, students enroll with a range of abilities and as such, there are students who do fail to make progress, can get stuck in remedial courses and may not know how to achieve their goals (Raby & Rhodes, 2018). While community college students face numerous barriers that impede overall success (Dougherty, 1994), policies and programs created in this century now show a greater chance of leading students to increased persistence (finishing the term) and completion (finishing their academic/professional program) (CCCSE, 2017) thus accounting for differential progression (Raby, 2012). Similar attention to retention and success is being adopted at the California State Universities as well. One such success-oriented program involves participating in on-line interactive educational programs that are designed to enhance the education abroad experience. On-line interactive programs have a proven record of engagement and ultimately student success.

This chapter focuses on the Center for Global Education Online Resource Courses (ORCs) that include numerous modules, tasks and action steps to support pre-departure, in-country, and re-entry learning affiliated with studying abroad. The ORCs are successful because they individualize instruction by using the core curriculum related to known ways in which to build critical thinking skills and enhance student learning, with a special focus on the international study abroad context. Through active learning, the ORCs not only enhance overall student comprehension, but also help to build a cohesive learning community with other students in similar programs. Finally, the ORC builds student engagement, enhances classroom learning, transcends to multi-college classroom dialogue, and builds technological skills that then enhance overall student success.

DIGITAL DIVIDE

The application of a digital divide began with the introduction of technology and yet, is most known regarding access to and use of computer and internet technology. Today, computer access is increasingly common, especially in connection to cell-phones, yet, there is still a deepening divide that is reflected in unequal access to information, uses for student learning, and application of technological skills in the workforce (Florida, 2017). Digitalization is embedded in contemporary society and has transformed the global economy with the creation of new jobs that demand very specific skills. Computer literacy is strongly correlated with employment security, overall wages, and higher levels of higher education (Muro, Liu, & Kulkarni, 2017). Differential skills use and access is negatively defined by educational stratification that varies across racial, class, gender and age and are intensified by intersectionality. Generation X and Generation Z students, who were born in the age of digital technology and are assumed to have digital competencies, are called Digital Natives. However, many of today's college students do not have those skills and when they enter higher education, they are at a disadvantage compared to their peers to actually have digital competencies (DeBruyckere, Kirschner, & Hulshof, 2016). Resulting stratification exists and contributes to under-representation in some college majors (Anderson, 2015). Women, African-Americans, and Latinos remain under-represented in the highest levels of digital employment (Florida, 2017). There are also geographic regions within highly digitalized states, such as California, that remain under-digitalized. For example, Stockton, Fresno, Bakersfield and Riverside are under-digitalized despite a California State University and number of community colleges that service those areas (Florida, 2017).

According to Florida (2017), there are three ways to minimize the digital divide. The Center for Global Education Online Resource Courses use all three of these techniques. First, on-line learning models provide known skills to enhance student knowledge of digitalization (Florida, 2017; Muro, Liu, & Kulkarni, 2017). The Center's ORCs use specifically developed on-line learning models to be interactive, to highlight knowledge about study abroad, to provide additional skill training in the use of computers that can be used to enhance overall Digital Literacy, to intensify student engagement and to result in student learning success. Secondly, there is a need to expand digital literacy to under-represented groups. In California, more women than men study abroad. There is also increasing diversity of students of color who also study abroad. The ORCs are created to target these groups and include specific learning modules with students of color in mind. In this context, the ORCs enhance knowledge for populations that need it the most. Thirdly, there is a need to not only teach "hard skills" but also "soft skills" including adaptability, emotional intelligence, and curiosity. While education abroad is known to provide these skills (Twombly, Salisbury, Tumanut, & Klute, 2012), the ORCs bridge the gap by highlighting soft skill learning into all modules. Finally, digitalization is stronger is specific urban digital hubs and absent in other geographical regions. The relationship that the Center has made with higher educational institutions throughout the state helps to bridge access to these geographic areas.

CENTER FOR GLOBAL EDUCATION ONLINE RESOURCES

The Center for Global Education is involved in research and resource development efforts as well as international education initiatives at the local, state, regional, national, and international levels. The resources are targeted for students, parents, and education abroad professionals (faculty and staff) to understand the purpose of education abroad with the *Study Abroad: Now More Than Ever Resources,* to refine search skills with *World Wide Colleges and Universities*, to introduce *Research on Education Aboard* resources, including project reports on *Barriers to Study Abroad in the California Community Colleges* and the *California Community College Student Outcomes Abroad Research (CCC SOAR) Project.* Finally, the resources include the *Safety Abroad First—Educational Travel Information (SAFETI) Clearinghouse* and the *Student Study Abroad Safety Handbook and AllAbroad.us: Diversity Outreach for Study Abroad.* The latter two are connected to the Online Resource Courses that will be the focus of this chapter.

Online Resource Courses

The Center houses three ORCs for education abroad, a) *GlobalScholar.us,* for U.S. Study Abroad Students, b) *uStudy.us,* for International Students Coming to the US, and c) *iStudent101.com,* for Any Student—Going from Any Country to Any Other Country. The purpose of these courses is for faculty or staff to provide access for students to use them independently or to incorporate them in a more formal orientation program or education abroad course to cover new material as well as to reinforce important information. In each on-line module, students are able to independently complete separate tasks through designed action steps. Completing all the tasks and action steps results in the completion of the module. Students, faculty, and staff can view the completion rate within the design of the online courses. There is also an opportunity for students to explore the ORC content for their own self-education. Each course includes a series of modules in which students evaluate their reasons for going abroad, what type of program they are looking for, where they would like to study, and what they want to achieve by doing so. Other modules include specific knowledge important to the education abroad process including understanding a country's culture, laws, customs, and politics. For students from diverse backgrounds, specific modules connect to resources in AllAbroad.us: Diversity Resources for Study Abroad with special information for students from diverse backgrounds and advice from study abroad students and professionals in the field developed to support study abroad by students from diverse ethnic backgrounds. Finally, specific modules work with the Student Study Abroad Safety Handbook (StudentsAbroad.com) to help students and their parents prepare for realities abroad and return home safely. Resources include comprehensive background information, useful checklists and questions, web resources, a sample emergency card and crisis response tools, and relevant words, phrases, and icons to help with international communication and assessment specific to each module. Course One (1) focuses on choosing a program and preparing for study abroad. Course Two (2) focuses on enhancing the experience during the student's time abroad. Course Three (3) focuses on the return home and enhancing the transition, providing outreach to others based on their experiences abroad, and continuing their international learning after the return home. Students who complete all three courses, including the Info Log assignments and the Community College or home-campus outreach presentations required in the third course, will earn the Global Scholar Certificate awarded by the Center and recognized by universities throughout the United States.

Global Scholar Online Courses. These ORCs introduce students to the opportunities and challenges inherent in participating in study abroad programs. Students learn tips and techniques to prepare for their program, how to cope with challenges that may arise while they're abroad, how to make the most of their study abroad experience while they're there, how to deal with issues that may arise after their return home, and advice for integrating their international learning after returning to their home campus, and career preparation. These ORCs include information from students who have studied abroad, staff who work with them in the U.S. and abroad, faculty who teach programs, and researchers who develop materials to help students through the process. Narratives from different ethnic, racial, and gender groups help to create accessibility to non-traditional students.

uStudy.us and iStudent101.com. These ORCs are designed to use the same framework as GlobalScholar.us but are intended for students from outside the U.S. who want to study in a U.S. college or university (uStudy.us), or come from a country outside the U.S. and study in a country outside the U.S., without being U.S.-centric (iStudent101.com), like the other ORCs. All courses are designed using an international curriculum to orient, train, and support students before, during, and after they study abroad, highlighting the health and safety concerns of institutions and students. They also recognize returning study abroad students who have completed the online curriculum and outreach through an International Honors Certificate Program.

DESIGNS WITH CAMPUS-BASED FACULTY
AND STAFF SUPPORT INTEGRATED

Along with students being able to access the content of each of these courses independently, all courses have been designed with the ability of any faculty or staff to be set up as an administrator for the course and be able to monitor the progress of students taking any of the courses in their group. Faculty and staff then also have an ability to choose which content to emphasize and support learning through connected group activities and/or interactions with individual students. Some faculty and staff only provide a link to the content with little or no connection. Others integrate it into their informal orientation and support programming. Others integrate it into a for-credit course or use it as the primary content for a for-credit course. With the inconsistency of training for faculty and staff in the study abroad field, there are others that may also not know about these ORCs or other available resources to enhance the study abroad experience.

DIGITAL DIVIDE AND EDUCATION ABROAD

Community college education abroad is grounded in the open access philosophy that allows admittance to educational programs for all students. Access is given to community members not enrolled as students, enrolled students, re-entry university students, and high school students who concurrently enroll in the college. As such, there is a wide range of students with different ability skills. Assessment of education abroad shows that students with the lowest pre-study abroad success in college who study abroad gain the highest levels of change towards their success in terms of retention, completion, and even moving out of remedial studies after study abroad (Raby, Rhodes, & Biscarra, 2014). Community college study abroad students represent a broad spectrum of gender, racial, ethnic, social class and ages (Raby, 2008). Community college research recognizes that different learners have unique educational pathways (Compton, Cox, & Laanan, 2006).

COHORT LEARNING

In community colleges, social integration is obtained by cohort involvement with on-campus programs (Gipson, Mitchell, & McLean, 2017) and with faculty (Wood & Harris, 2015). Most community college education abroad classes/programs are faculty-led in which faculty design, lead, and market programs, often, to their own students with whom they have developed special bonds. Student cohorts are critical to the success of student learning as students maintain special bonds with their peers, which enhances engagement. This is due, in part, to adult learners being best able to relate to other adults because of similar backgrounds and experiences (Zhang, Lui & Hagedorn, 2013). Student cohorts remain important as peers encourage each other to study abroad and provide a familiarity important for building encouragement, comfort, and safety (Amani & Kim, 2017; Brenner, 2016; Willis 2016).

NON-TRADITIONAL STUDENTS

California Community College and California State University students are inherently diverse. Likewise, the students from these institutions who enroll in education abroad programs are also diverse (IIE, 2017). The education abroad field is focusing on increasing the percentages of diverse racial and ethnic groups who study abroad. Progress is being made to move towards matching their overall percentage of enrollment in higher education and the racial and ethnic percentages of students who study abroad, nationally and at the California state level. There are more women who study abroad than men, more adults who study abroad with the average age of community college students being 28 years old (Raby & Rhodes, 2018). These are the same demographics as those who are negatively influenced by the digital divide.

DIGITAL APPLICATIONS

Education abroad requires some digital competencies. In a survey with California community college education abroad directors, several noted that many students have difficulty with technology during the application process and that extra assistance and flexibility is needed to help set the tone for a successful learning experience (Raby & Rhodes, 2018). In the same survey, respondents agreed that despite the advances of social media and the fact that so many of the students are expected to be digital natives, there is limited success in the use of on-line marketing and outreach for community college students. While this may be different that the general search of community college information, for study abroad, students prefer face-to-face interactions as they may not have previous information about study abroad and see the direct connection to faculty, staff and students who have studied abroad to be critical to overcoming resistance to seeing study abroad as a safe and useful opportunity that is relevant and available for them (Raby & Rhodes, 2018). While many colleges list their programs on the California Colleges for International Education Clearinghouse Web page (www.ccieworld.org), they may not use other social media outlets. 12% said they update their campus Web page more than once a year. Only four respondents use Facebook to share information on financial aid and scholarships. The digital divide is clearly shown in the minimal use of on-line technologies in the community college education abroad world

ENGAGEMENT AND STUDENT SUCCESS

Academic engagement is a critical component in the learning process that contributes to academic success. The more engaged the student is to their studies, the greater the potential for success in terms of learning and development goals associated with persistence and completion (Astin, 1984; Booth, et al., 2013; Kuh, et al., 2006; Tinto, 1993). Academically engaged students are more dedicated to their studies, stay in school, and have a better chance of completion. Colleges that establish programs to increase academic engagement will improve the odds of student success (Tinto, 2010). Engagement best occurs when planned academic (in-class) and co-curricular (out-of class) activities intersect and is often found in high-impact educational activities. Education abroad is a high-impact educational activity (Chieffo 2010; Sutton & Rubin 2010).

ENGAGEMENT ACTIVITIES

Engagement intensifies when specific characteristics are met that inspire student-initiated motivation (Chickering & Kuh, 2005). Some of these characteristics include a) deepened commitment through purposeful tasks; b) extended and substantive interactions with faculty and peers; c) frequent feedback to student performance; d) interactions with people who are different than themselves; e) application of what students learn in different settings; and f) life changing experiences. These characteristics are often found in such college activities such as first-year seminars, internships, learning communities, service-learning/community-based learning, writing intensive courses, collaborative projects and assignments, capstone projects and diversity/global learning (Kuh, Kinzie, Schuh, Whitt, & Associates, 2005). The potential for change is enhanced through the residential experience that is integrated into most study abroad programs, but may not be available at a community college, as most community colleges do not own or manage housing for their students. The ORCs adapt these high impact practices as part of most study abroad program design and thus capture activities that lead to student engagement.

ON-LINE LEARNING AND STUDENT ENGAGEMENT

On-line communities and learning are also a noted activity that inspires student motivation and enhances engagement (Chickering & Kuh, 2005) and are one way in which to create high-impact educational activities (Clothey & Austin-Li, 2009). It is the collaboration in these on-line communities that provide formal and informal learning opportunities that critically engage the learner in the learning process (Baskerville, 2012; Dangle & Wang, 2008; Mokena, 2013; Wankel & Blessinger, 2012). Use of on-line learning communities is an accepted method to enhance learning opportunities in the community college classroom (Johnson et al., 2013; Ryland, 2013). The ORCs are designed to bridge in-class and out-of-class learning modalities.

STUDY ABROAD AND STUDENT ENGAGEMENT

Harper and Quaye (2009) note that the "participation in educationally effective practices, both inside and outside the classroom, leads to a range of measurable outcomes" (p. 3). Education abroad is one of these educationally effective practices. Many community college programs that promote positive outcomes involve close interaction with faculty and culturally diverse groups of people in hands-on situations (McCormick, Kinzie, & Gonyea, 2013). Similarly, community college education abroad involves close interaction with faculty, focuses on student cohorts, uses reflection as a learning activity, and when managed properly can be a way to help students develop in line with institutional goals and learning outcomes (Raby, Rhodes, & Biscarra, 2014). The student engagement literature found a link between student background in terms of age, ability, and effort toward college persistence and completion (McCormick, Kinzie, & Gonyea, 2013). Yet, for community college education abroad, a student's background is not a factor because interest rather than age or ability is the main criterion for enrollment.

Student engagement in international and intercultural learning can lead to improvement in overall success and can result in changed values, life-goals, and eventually in personal expectations. When students face inter-cultural challenges while abroad, they are forced to make complex choices that prepare them to make similar complex choices later in life (Blair, 2013). There is empirical evidence of the positive effects of education abroad that result in distinct changes in students in terms of sociability, cultural identity development, ethno-relativism development, inter-group tolerance, and global-mindedness changes (Bolen 2007; Forum 2009; Hammer, Bennett, & Wiseman 2003; Paige et al., 2009; Sutton, Miller, & Rubin, 2007; Sutton & Rubin 2004; University of Minnesota-Sage Project 2009). There is noted growth in acquisition of international literacy skills (Amani, 2011; Willis, 2016), knowledge of the global economy (Niser, 2010), citizenship skills (Frost & Raby, 2009), and intercultural awareness skills (Emert & Pearson, 2007). Community colleges use strategies (Stovall, 2001) that promote increased contact with peers and faculty that result in successful social and academic integration. On-line learning communities that combine on-line and in-person are increasingly part of Global Certificate programs (Rodriguez, 2016). Raby, Rhodes, and Biscarra (2014) concluded from their research in the CCC-SOAR project of community colleges that there are three distinct ways in which education abroad leads to student success. First, students gained content specific information that was defined by the courses that they took. Students also gained country specific information in terms of the particular issues unique to that country and special country-specific health and safety issues. Secondly, specific activities offered prior to going abroad, during the education abroad program and after the students returned to campus helped to heighten student learning. Finally, student reflections on their experiences indicated a tremendous range of international learning. Among the most valuable can be learning how a different culture functions and the use of language in interactions with locals. In terms of the impact, students wrote that education abroad gave them a worldly view on life, enabled them to learn about both content and culture, which came from being in another country as well as from the group experience.

DECREASING THE DIGITAL DIVIDE WITH ONLINE RESOURCES

There are many ways in which online resources when used with education abroad can help to decrease the digital divide.

Commitment through Purposeful Tasks

The Center's Online Resource Courses enhance student commitment to learning as a result of two purposeful tasks. The first task is acquisition of computer literacy skills that support larger educational and career goals. Technology and social media become a directed tool for class work to link reading, writing, and comprehension literacy to critical thinking skills needed in work-environments. In this process, students learn to take charge of technology in the learning process, rather than to have technology control the student (Raby & Kaufman, 2000). Subject-content specific sites are combined with general informational sites, digital storytelling sites, journalistic sites, and a range of social media that introduces current information that is then filtered through a variety of viewpoints. This is important since integration of technology improves student outcomes (Natow, Reddy, & Grant, 2017). As an instructional technology, the ORCs provide content as well as assessment mechanisms and opportunities for students to reflect and apply the material. The second purposeful task is the development of factual knowledge and understanding of that knowledge that uses critical thinking skills. Students not only learn academic content particular to the individual course, but they learn about it from the perspective of multiple countries and in multiple contexts. In that students like educational video games and the use leads to success (Martí-Parreño, Galbis-Córdova, & Miquel-Romero, 2018), the impact of ORCs becomes even more important.

Extended and Substantive Interactions With Faculty and Peers

Interactions between student-faculty and student-peers allow students to construct new knowledge via social interaction which then allows them to take an active role in effective conceptual change and transformative learning constructs (Orey, 2008; Wankel & Blessinger, 2012). There are opportunities for extended and substantive interactions between students and their peers as well as with faculty. A basic principle of internationalizing curricula is the acquisition of skills to perceive multiple perspectives, to reconcile conflicting ideologies, and to respect a relativity of differences. This is what Green (2012) refers to as an "international mind-set." In order to achieve this literacy, curricula must include context for active learning in which one is conscious of different meanings and changing expressions that are governed by cultural differences. Internationalization in this context allows an individual to change their focus to say things like: "I never thought of that," or "I never knew I could apply it that way" (Fersh, 1993, p. 7). While internationalization can occur through a range of activities, such learning is often associated with studying abroad in which students travel to learn (Knight 2004). However, similar pedagogical learning experiences can also occur while the student remains in the classroom without any travel through on-line discussion groups (Roller, 2012). The ORCs intersect both in-classroom and out-of-classroom learning.

Application of What Students Learn in Different Settings

The contemporary focus of the themes explored in ORCs provide real-life information that students can apply in different settings. The interconnection of multiple academic and professional perspectives is the cornerstone. This construct is consistent with the belief that a truly educated citizen needs to understand that all issues intersect all aspects of community building and local employment. The multi-disciplinary focus furthermore underscores that international issues are not the purview of any one part of the curriculum but result from the integration into all community college curricula. Learning in different settings benefits students as they get a holistic understanding of contemporary issues. Moreover, it is this holistic understanding that facilitates the critical engagement of the learner in the learning process (Baskerville, 2012). The realization that decisions can not be made in isolation as well as the global influences across disciplines, changes not only students' perceptions, but affects the manner in which they look at other students and members of their community (Green, 2012; Kaufman, 1998).

FUTURE AND EMERGING TRENDS

Simulation through the integration of technology is an emerging trend (Johnson et al, 2013). Another trend in higher education is a focus on student outcomes and learning assessment. Literature confirms the importance of using technology to engage the learner in the learning process (Wankel & Blessinger, 2012). ORC is an identified way to document student performance and engagement levels. Student assessment takes two forms: student learning and student engagement. Student learning is contingent on the multi-disciplinary construct, which changes to meet the needs of college students. Thus, while some students respond more to the reading and writing assignments, others respond to the research they must do to respond to questions and reflection exercises, which are at the core of the ORCs. Another emerging trend is the use of student engagement to support college persistence and completion. As noted, the more students are engaged, the stronger the potential learning (Booth, et al., 2013). Attendance is one element of student engagement that has broad-ranging repercussions. In economically challenging times, the emphasis on student persistence draws attention to the potential for ORCs to maximize the impact of study abroad, beyond their in-class curricular learning..

CONCLUSION

For the past two decades, the Center's Online Resource Courses have been an innovative approach to teaching and learning. Evaluations show that the Online Resource Courses have a noted effect on the construction of knowledge, increased student engagement, increased use of technology, all of which lead to overall student success. The spiraled activities within each ORC enforce the range of positive outcomes that occur when students are directly engaged in their learning. Finally, the ORCs remain a valuable conduit for internationalization of the curriculum, which provides critical learning skills for a global economy.

Currently, few universities and colleges offer courses and long-term orientation programs for education abroad and for international students studying in the U.S., or in other countries for students moving from a non-US country to a non-US country to support these students before, during, and after the education abroad experience. Some of the colleges and universities who have implemented a comprehensive orientation program have done so with the support of the ORCs described in this chapter.

Access to high-quality education is unequal and over time those inequalities build on themselves. Community colleges have contributed to this problem, but they are also essential to the solution (Bailey, 2012). It is known that community college students face numerous barriers both externally and internally that can impede them from achieving their goals for success. The use of the Center's Online Resource Courses combined with education abroad take two high level educational programs and use them to positively effect student learning and in the process contribute to countering inequalities by providing unique high impact opportunities.

It is important to note that the impact from education abroad alone or on-line learning alone does not necessarily result in the most enhanced and impactful levels of student learning. However, it is the intentional integration of the two that can help maximize the impact on retention and success and other positive outcomes of study abroad. As such, knowing the importance of intent, the link to success can become even more apparent if community college education abroad programs begin to intentionally offer on-line learning components into their program to facilitate meaningful academic and personal growth through purposeful design.

There is a pressing demand for re-focusing the delivery of curriculum and programs that express innovation and which define skills that students need and want. A critical question is whether or not community colleges are even beginning to address those skills in a systematic and coordinated manner. Such an undertaking begins with the critical examination of college policies, programs, and practices. Focus needs to remain on whether or not the college is improving or expanding international literacy skills that impact future job attainment and personal growth. The primary issue is the ability of visionary community college leadership to understand and make changes at the colleges so that their students have the competitive edge to "integrate and apply their academic, technical and practical knowledge and skills to solve real-world problems, to continue learning in formal and informal ways throughout their lifetimes on-the-job, in schools and in their communities, and to work effectively with other people as customers, coworkers and supervisors" (Porter, 2002, p. 4). Since internationalization has always been more than just meeting employer's needs, the discussion needs to begin with the understanding that global flows are simply part of our lives. Study abroad can be the highlight of a community college students' learning, impacting their retention and success and the direction of their future lives. Along with providing additional study abroad opportunities for students, providing enhanced learning before, during, and after study abroad is critical for maximizing the potential that study abroad can have for improving the learning and retention and success of community college students.

REFERENCES

Amani, M. (2011). *Study abroad decision and participation at community colleges: Influential factors and challenges from the voices of students and coordinators* (Doctoral dissertation). Available from ProQuest Dissertations and Theses database. (UMI No. 3438831)

Amani, M., & Kim, M. M. (2017): Study abroad participation at community colleges: Students' decision and influential factors. *Community College Journal of Research and Practice, 41*(10), 1–5.

Anderson, M. (2015). *Racial and ethnic differences in how people use mobile technology.* Retrieved from http://www.pewresearch.org/fact-tank/2015/04/30/racial-and-ethnic-differences-in-how- people-use-mobile-technology/

Astin, A. W. (1984). Student involvement: A developmental theory for higher education. *Journal of College Student Personnel, 25,* 297–308.

Bailey, T. R. (2012). Equity and community colleges. *The Chronicle Review.* Retrieved from, http://chronicle.com/article/EquityCommunity-Colleges/132643/

Baskerville, D. (2012). Integrating on-line technology into teaching activities to enhance student and teacher learning in a New Zealand primary school. *Technology, Pedagogy and Education, 21*(1), 119–35.

Blair, S. G. (2013). *Using external assessment instruments sensibly in education abroad: The best assessment strategies are the most diversified ones.* Retrieved from, www. GoWithCEA.com/Assessment

Bolen, M. (2007). *Changing perspectives in international education.* Bloomington: IN: Indiana University Press.

Booth, K., Cooper, D., Karandjeff, K., Large, M., Pellegrin, N., Purnell, R., Rodriquez-Kiino, D., Schiorring, E., & Willett, T. (2013). *Using student voices to redefine support: What community college students say institutions, instructors and others can do to help them succeed.* Berkeley, CA: The Research and Planning Group for California Community Colleges.

Brenner, A. (2016). Transformative learning through education abroad: A case study of a community college program. In R. R. Latiner & E. J. Valeau (Eds.), *International education at community colleges: Themes, practices, research, and case studies* (pp. 370–390). New York, NY: Palgrave.

California Colleges for International Education. (n.d.). *Clearinghouse web page.* Retrieved from www.ccieworld.org.

Center for Community College Student Engagement. (2017). *The community college student report.* Retrieved from, http://www.ccsse.org/aboutsurvey/biblio/page1.cfm

Center for Global Education. (n.d.). *Student study abroad safety handbook.* Retrieved from www.StudentsAbroad.com

Center for Global Education. (n.d.) UStudy. us. www.uStudy.us.com.

Center for Global Education (n.d.). iStudent101. www.iStudent101.com

Chickering, A. W., & Kuh, G.D. (2005). *Promoting student success: Creating conditions so every student can learn.* Retrieved from, http://nsse.iub.edu/_/?cid=128

Chieffo, L. (2010). *The freshman factor: Outcomes of short-term education abroad programs on first-year students.* Newark, DE: Center for International Studies, University of Delaware. http://www.udel.edu/global/pdf/freshmen-abroad-outcomes.pdf

Clothey, R., & Austin-Li, S. (2009). Building a sense of community through online video. *International Journal of Web-Based Communities, 6*(3), 303–316.

Compton, J., Cox, E., & Laanan, F. (2006). Adult learners in transition. *New Directions for Student Services, 114*, 73–80. doi:10.1002/ss.208

Dangle, H., & Wang, C. (2008). Student response systems in higher education: Moving beyond linear teaching and surface learning. *Journal of Educational technology Development and Exchange, 1*(1), 93–104.

De Bruyckere, P., Kirschner, P. A., & Hulshof, C. D. (2016). *Technology in education: What teachers should know.* Retrieved from https://www.aft.org/ae/spring2016/debruyckere-kirschner-and-hulshof

Dougherty, K. J. (1994). *The contradictory college: The conflict origins, impacts, and futures of the community college.* Albany, NY: State University of New York Press.

Emert, H. A., & Pearson, D. L. (2007). Expanding the vision of international education: Collaboration, assessment, and intercultural development. In Valeau, E. J. & Raby, R. L. (Eds.), *International reform efforts and challenges in community colleges.* San Francisco, CA: Jossey-Bass.

Fersh, S. (1993). *Integrating the trans-national/cultural dimension. Fastback # 361: No. 104.* Bloomington, Indiana: Phi Delta Kappa Educational Foundation.

Florida, R. (2017). *America's digitalization divide.* Retrieved from, www.citylab.com/equity/2017/11/americas-digitalization-divide/546080/

Forum on Education Abroad. (2009). *Forum BEVI Project.* Retrieved from http://www.forumea.org/research-bevi.htm.

Frost, R. A., & Raby, R. L. (2009). Democratizing study abroad. In R. Lewin (Ed.), *The handbook of practice and research in study abroad: Higher education and the quest for global citizenship* (p. 586). New York: NY: Routledge.

Gipson, J., Donald Jr., M., & McLean, C. (2017). An investigation of high-achieving African-American students attending community colleges: A mixed methods research study. *Community College Journal of Research and Practice, 42*, 1–13.

Green, M. F. (2012). *Measuring and assessing internationalization.* Washington, DC: NAFSA Publications. Retrieved from, http://www.nafsa.org/Resource_Library_Assets/Publications_Library/Measuring_and_As sessing_Internationalization/

Hammer, M. R., Bennett, M. J., & Wiseman, R. (2003). Measuring intercultural sensitivity: The intercultural development inventory. *International Journal of Intercultural Relations, 27*(4), 421–443.

Harper, S. R., & Quaye, S. J. (2009). Beyond sameness, with engagement and outcomes for all: An introduction. In S. R. Harper & S. J. Quaye (Eds.), *Student engagement in higher education: Theoretical perspectives and practical approaches for diverse populations* (pp. 1–15). New York, NY: Routledge.

Institute of International Education. (2016). *Open Doors 2017 data tables: Community college demographics*. Retrieved from, www.iie.org/en/Research-and-Insights/Open-Doors/Data/Community-College-Data-Resource

Johnson, L., Adams Becker, S., Cummins, M., Estrada, V., Freeman, A., & Ludgate, H. (2013). *Technology outlook for community, technical, and junior colleges 2013–2018: An NMC Horizon Project Sector Analysis*. Austin, TX: The New Media Consortium. Retrieved from http://www.nmc.org/pdf/2013-technology-outlook-community-colleges.pdf

Kaufman, J. P. (1998). Using simulation as a tool to teach about international negotiation. *International Negotiation 3(*1), 5–75.

Knight, J. (2004). Internationalization remodeled: Definition, approaches, and rationales. *Journal of Studies in International Education, 8*(1), 5–31.

Kuh, G. D., Kinzie, J., Schuh, J. H., Whitt, E. J., & Associates. (2005). *Student success in college: Creating conditions that matter*. San Francisco: CA: Jossey-Bass.

McCormick, A. C., Kinzie, J., & Gonyea, R. M. (2013). Student engagement: Bridging research and practice to improve the quality of undergraduate education. In Paulsen, M. B.(Ed.), *Higher education handbook of theory and research* (pp. 47–92). Berlin: Springer.

Martí-Parreño, J., Galbis-Córdova, A., & Miquel-Romero, M.J. (2018). Students' attitude towards the use of educational video games to develop competencies. *Computers in Human Behavior, 81*, 366–377.

Mokoena, S. (2013). Engagement with and participation in online discussion Forums. *The Turkish Online Journal of Educational Technology, 12*(2), 1–9.

Muro, M., Sifan, L., & Siddharth, K. (2017). *Digitalization and the American workforce*. New York, NY: Brooking's Metropolitan Policy Program. Retrieved from www.brookings.edu/wp-content/uploads/2017/11/mpp_2017nov15_digitalization_appendix_a.pdf

Natow, R. N., Reddy, V., & Grant, M. (2017). *How and why higher education institutions use technology in developmental education programming*. Retrieved from, https://postsecondaryreadiness.org/how-why-higher-education-institutions-use-technology-developmental-education-programming/

Niser, J. C. (2010). Study abroad education in New England higher education: A pilot survey. *International Journal of Educational Management, 24*(1), 48–55.

Orey, M. (2008). *Emerging perspectives on learning, teaching and technology*. Bloomington, IN: Association for Educational Communications and Technology.

Paige, R. M., Fry, G., LaBrack, B., Stallman, E. M., Josić, J., & Jon, J.-E. (2009). *Study abroad for global engagement: Results that inform research and policy agendas*. Paper presented at the annual conference of the Forum on Education Abroad, Portland, OR. February 22, 2009.

Porter, M. E. (2002). *GWIT.* Paper presented at the Inter-American Development Bank, Washington, DC. November 18.

Raby, R. L. (2008). *Meeting America's global education challenge*: *Expanding education abroad at U.S. community colleges*. New York, NY: Institute for International Education.

Raby, R. L. (2012). Re-imagining international education at community colleges. *Audem: International Journal of Higher Education and Democracy, 3*, 81–99.

Raby, R. L., & Kaufman, J. (2000). The international negotiation modules project: Using computer-assisted simulation to enhance teaching and learning strategies in the community college. In L. A. Petrides (Ed.), *Cases of information technology in higher education.* Hershey, PA: Idea Group Publishing.

Raby, R. L., & Rhodes, G. M. (2018). Promoting education abroad among community college students: Overcoming obstacles and developing inclusive practices. In H. B. Hamir & N. Gozik (Eds.), *Promoting inclusion in education abroad* (pp. 114–133). London: Stylus.

Raby, R. L., Rhodes, G. M., & Biscarra, A. (2014). Community college study abroad: Implications for student success. *Community College Journal of Research and Practice, 38,* 174–183.

Rodriguez, P. (2016). Global certificates: Bringing intentionality and ownership to comprehensive internationalization. In R. R. Latiner & E. J. Valeau (Eds.), *International education at community colleges: Themes, practices, research, and case studies* (pp. 283–299). New York, NY: Palgrave.

Roller, K. M. (2012). *Pre-Service teachers and study abroad: A reflective, experiential sojourn to increase intercultural competence and translate the experience into culturally relevant pedagogy.* Unpublished doctoral dissertation, University of California at Los Angeles.

Ross-Gordon, J. M. M. (2011). *Research on adult learners: Supporting the needs of a student population that is no longer nontraditional (returning adult students).* Retrieved from, https://www.aacu.org/publications-research/periodicals/research-adult-learners- supporting-needs-student-population-no

Ryland, J. N. (2013). *Technology and the future of the community college.* American Association for Community College. Retrieved from, www.aacc.nche.edu/Resources/aaccprograms/pastprojects/Pages/technologyfuture.aspx

Stovall, M. (2001). Using success courses for promoting persistence and completion. In S. Aragon (Ed.) *Beyond access—Methods and models for increasing retention and learning among minority students new directions for community colleges* (pp. 45–54). San Francisco, CA: Jossey Bass.

Sutton, R., Miller, A., & Rubin, D. (2007). *A guide to outcomes assessment in education abroad.* Carlisle, PA: The Forum for Education Abroad.

Sutton, R. C., & Rubin, D. L. (2004). The GLOSSARI project: Initial findings from a system-wide research initiative on study abroad learning outcomes. *Frontiers: The Interdisciplinary Journal of International Education, 10,* 65–82.

Sutton, R. C., & Rubin, D. L. (2010). *Documenting the academic impact of study abroad: Final report of the GLOSSARI project.* Paper presented at the annual conference of NAFSA: Association of International Educators, Kansas City, MO. May.

Tinto, V. (1993). *Leaving college: Rethinking the causes and cures of student attrition* (2nd ed.). Chicago, IL: University of Chicago Press.

Tinto, V. (2010). From theory to action: Exploring the institutional conditions for student retention. *Higher Education: Handbook of Theory and Research, 25*(2), 51–89.

Twombly, S. B., Salisbury, M. H., Tumanut, S. D., & Klute, P. (2012). Study abroad in a new global century—Renewing the promise, refining the purpose. *ASHE Higher Education Report, 38*(4), 1–152.

University of Minnesota Sage Project. (2009). *Study abroad/non-study abroad graduation rates by college.* Twin Cities, MN: Office of Institutional Research, University of Minnesota—Twin Cities. Retrieved from, www.umabroad.umn.edu/assets/files/PDFs/ci/Evaluation %20Pages/graduationRatesbyCollege.pdf

Wankel, C., & Blessinger, P. (2012). *Increasing student engagement and retention using classroom technologies: Classroom response systems and mediated discourse technologies.* Bingly UK: Emerald Publishing.

Willis, T. Y. (2016). Microaggressions and intersectionality in the experiences of Black women studying abroad through community colleges: Implications for practice. In R. L. Raby & E. J. Valeau (Eds.), *International education at community colleges: Themes, practices, research, and case studies* (pp. 167–186). New York, NY: Palgrave.

Wood, J. L., & Harris, F., III. (2015). The effect of academic engagement on sense of belonging: A hierarchical, multilevel analysis of Black men in the community colleges. *Spectrum: A Journal on Black Men, 4*(1), 21–47

Zhang, Y. L., Lui, J., & Hagedorn, L. S. (2013). Post transfer experiences: Adult undergraduate students at a research university. *Journal of Applied Research in the Community College, 21,* 31–40.

BIOGRAPHIES

Dr. Jeffrey S. Brooks is Associate Dean for Research and Innovation and Professor of Educational Leadership in the School of Education at the Royal Melbourne Institute of Technology University (RMIT), based in Bundoora, Australia. He is a two-time J. William Fulbright Senior Scholar alumnus who has conducted studies in the United States, Australia, Thailand, Indonesia and the Philippines and in many other cross-national contexts. His research focuses broadly on educational leadership, and he examines the way leaders influence (and are influenced by) dynamics such as racism, globalization, social justice, student learning and school reform. Dr. Brooks is author of over 100 scholarly publications and he has been a leader and team member in projects that have garnered over 8 million dollars extramural funding. He is author of two full-length books based on his research: *The Dark Side of School Reform: Teaching in the Space between Reality and Utopia* and 2013 AESA Critics Choice Award-winning *Black School, White School: Racism and Educational (Mis)leadership*. He is co-author of *Foundations of Educational Leadership: Developing Excellent and Equitable Schools* with Anthony H. Normore. He is also co-editor of twelve volumes, including *Leading Against the Grain: Lessons for Creating Just and Equitable Schools*. Dr. Brooks has written many articles in leading refereed educational research journals, including *Educational Administration Quarterly*, *Educational Policy*, *Journal of Educational Administration*, and *International Journal of Educational Management*. Dr. Brooks is Series Editor for the *Educational Leadership for Social Justice* book series

Crossing the Bridge of the Digital Divide: A Walk with Global Leaders, pages 227–237.
Copyright © 2019 by Information Age Publishing
227

and co-Editor of the *Journal of Educational Administration and History*. He has served in several leadership positions in universities and educational research organizations, AERA Division A (Administration, Organizations and Leadership) and the AERA Leadership for Social Justice Special Interest Group. He is the current Convener of the AARE Educational Leadership SIG.

Dr. Cheryl Brown is Associate Professor of e-learning and co-Director of the e-Learning Lab in the School of Education Studies and Leadership at the University of Canterbury, New Zealand. Previously she worked in the Centre for Innovation in Learning and Teaching at the University of Cape Town where she has managed e-learning projects on digital education leadership, personal mobile devices in learning and teaching and developing e-learning professionals in Africa. Her research interests are centred around access to ICTs and how they facilitate or inhibit students' participation in learning. In the past few years she has looked more closely at the role technological devices (for example cell phones and tablets) play in students learning in a developing context and the development of students' digital literacy practices.

Ruben Caputo has been working in Education Technology for the past 7 years at California State University Dominguez Hills. After graduating from Biola University with a Bachelor's of Science degree in Business Marketing, he obtained a position working for Apple's Education Team as a Partner Manager focused on Business development. It was during his time at Apple that he grew fond of working in the Education world. In 2011, he started implementing graduate level course curriculum design to the Dominguez Hills School Leadership program. Furthering his educational background, Ruben obtained his Master's degree in Technology Entrepreneurship from the University of Maryland. He is part of Google's EdTech team that promotes universal design practices to bring forth a more connected space for K–12 educators and learners. Ruben thrives to empower educators. He believes in equal access in order to personalize and bring engagement in education as one of the ways to better the world.

Dr. Laura Czerniewicz is the Director of the Centre for Innovation in Learning and Teaching (CILT) at the University of Cape Town in South Africa. She is an associate professor in the Centre for Higher Education Development, committed to equity of access and success in higher education. Her research interests include the technologically-mediated practices of students and academics, the nature of the changing higher education environment and the geopolitics of knowledge, underpinned by a commitment to surfacing the expressions of inequality within and across contexts. Dr. Czerniewicz is presently leading a project on the Unbundled University: Researching emerging models in an unequal landscape (http://unbundleduni.com/) together with colleagues at Leeds University. Laura is involved with policy work, is a contributor to national and global conversations in varied formats and serves on the advisory boards of a variety of international higher education educational and technology publications. Much of her work is online at https://uct.academia.edu/LauraCzerniewicz. She blogs intermittently and can be followed on Twitter as @czernie.

Dr. Xuefei (Nancy) Deng is an Associate Professor of Information Systems at California State University, Dominguez Hills. Dr. Deng's research interests include crowdsourcing, digital and social media, human value and IS design, IT workforce, and knowledge management. Her research has been published in the *MIS Quarterly, Journal of Management Information Systems, Journal of the Association for Information Systems, Decision Support Systems,* and *Information Systems Journal*, among others. She co-chairs "Digital and Social Media in Enterprise" mini-track at the Hawaii International Conference on Information Sciences and serves as Associate Editor for *Information and Organization,* and *Journal of Organizational Computing and Electronic Commerce.*

Dr. Rebecca Eynon is an Associate Professor and Senior Research Fellow at the University of Oxford. At the University she holds a joint academic post between the Oxford Internet Institute (OII) and the Department of Education. Since 2000 her research has focused on education, technology and inequalities, and she has carried out projects in a range of settings and life stages.

Dr. Yesenia Fernández is Assistant Professor of School Leadership, Cal State Dominguez Hills, is a first generation college student who earned a Ph.D. in Education at Claremont Graduate University. She studies how systemic racism in the public school system precludes students of color from higher education as well as the experiences of first generation college students. She has been student advocate for twenty years and recently served as an urban school district leader developing systems to improve equity and access to higher education. She is dedicated to working with school leaders and policymakers to ensure equity and justice in urban schools

Dr. Kitty Fortner is an Assistant Professor at California State University, Dominguez Hills in the College of Education Graduate Education Department working in the School Leadership Program. She received her doctorate in Leadership for Educational Justice from the University of Redlands in Redlands, California. She has been involved in K-12 urban education for over twenty years in public and public charter schools. She began her educational career as an elementary classroom teacher, working with low income, second language students. She has served as Principal, BTSA Support Provider, Technology Facilitator, RTI Coordinator, SPED Admin Designee and as a Master/Mentor Teacher assisting new and veteran teachers. She was founding principal at two successful charter schools. Her work with school leaders, teachers, pre-service teachers, school districts, and policymakers is aimed at assisting school leaders. Her passion is to provide school leaders with the tools that they need to transform their schools into innovative and excellent schools that provide students with the best learning experience possible. She believes that leadership plays an imperative role in the fostering of academic success for all learners. She is a co-author of a book entitled *Who we are and How we learn: Educational Justice for Diverse Learners* (2015, Cognella Publishing)

Dr. Anne Geniets is a Research Fellow with the Learning and New Technologies Research Group at the Department of Education, University of Oxford. She is a developmental psychologist and communications & media scholar, with a research focus on health, social justice, training and technology in low resource settings.

Dr. Tina N. Hohlfeld is a retired college professor and K–12 teacher with over 30-years of experience. Dr. Hohlfeld's research focuses on examining the relationship between technology integration and positive outcomes for K–12 students, especially students who are at-risk (e.g., Digital Divide); identifying best instructional practices for web-based distance learning; and delineating efficient methods for supporting both pre-service and in-service teachers as they learn to integrate technology into their daily instructional practices. Dr. Hohlfeld has published in multiple leading venues, including *Computers & Education, Educational Technology Research and Development*, and the *Journal of Research on Technology in Education.*

Dr. Julie Jhun is currently Assistant Professor of school leadership at California State University Dominguez Hills. Dr. Jhun began her career in education as a Teach for America corps member teaching in Lynwood, California. A strong advocate for positive behavioral reinforcement, her doctoral dissertation studied the life-changing effects of effective parenting techniques on adolescents. Dr. Jhun received her doctorate in education from the University of California, Los Angeles, her master's degree in education from Loyola Marymount University, and her bachelor's degrees in psychology and piano performance from Oberlin College and the Oberlin Conservatory of Music. A lifelong Angeleno, she resides with her husband and two children.

Jenell A. Krishnan is a former high school English Language Arts teacher with experience using instructional technologies to support her ultimate student learning goal—college and career readiness. She left her teaching career in New York State to pursue a doctoral degree at the University of California at Irvine's School of Education, specializing in Language, Literacy, and Technology. Her research interests lie at the interface of digital technologies, student literacy development, and teacher professional development. She has supported an IES-funded, digital literacy intervention, facilitated numerous teacher professional developments opportunities, and has published in *Teaching English with Technology* and *Journal of Child Language.*

Dr. Saili S. Kulkarni is an Assistant Professor of Special Education at San José State University. Dr. Kulkarni's research looks at the intersections of disability, race, and teacher education using a disability studies lens. She has authored several articles and chapters highlighting the perspectives of culturally and linguistically diverse students with disabilities and their families as well as the restructuring of special education teacher education to incorporate a disability studies and critical race theory framework (DisCrit). Dr. Kulkarni formerly worked for the Oakland Unified School district as an inclusive elementary school teacher. She completed her doctorate at the University of Wisconsin-Madison in Special Education with the support of the Arvil S. Barr Teacher Education Fellowship and as an Edward Alexander Bouchet Honor Society Member. At University of Wisconsin-Madison, Dr. Kulkarni was the program coordinator for the Multicultural Graduate Network, a program set to increase retention, recruitment, and support for graduate students of color on campus. Her aim is to make disability studies and advocacy more visible in mainstream contexts.

Dr. Antonia Issa Lahera currently works as an Associate Professor at CSU Dominguez Hills in the School Leadership Program. During her nearly 30 years in the field she has worked in urban settings as a teacher, staff developer and site principal. She has worked in highly innovative settings as the leader of a reconstituted school and also a school ending the social promotion of students to high school. Dr. Issa Lahera received her doctorate from the University of Southern California in 2003 in urban leadership. She has worked as a mentor for the National Urban Alliance and done extensive work around United States. In addition to directing 4 multimillion dollar federal grants for school improvement in the Los Angeles area Dr. Issa Lahera is the co-editor of two books including *Restorative practice meets social justice: Un-silencing the voices of "at-promise" student populations* (2017, Information Age Publishing) and *Pathways to excellence: Developing and cultivating leaders for the classroom and beyond'* (2014, Emerald).

Dr. Roslind Latiner is a Senior Lecturer at California State University, Northridge in the Educational Leadership and Policy Studies Department of the Michael D. Eisner College of Education. She is an affiliate faculty for the ELPS Ed.D. Community College program and serves as the Director of California Colleges for International Education, a non-profit consortium whose membership includes ninety-one California community colleges. Dr. Raby is also a lead faculty in the College of Humanities and Sciences, at the University of Phoenix, Southern California Campus. She received her Ph.D. in the field of Comparative and International Education from UCLA and since 1984, has worked with community college faculty and administrators to help them internationalize their campuses. Dr. Raby has been publishing in the field of community college internationalization since 1985. Among her 150 publications on the topic of international education and community colleges are: *Handbook of Comparative Studies on Community Colleges & Global Counterparts* (Springer, 2018); *International Education* *at Community Colleges: Themes, Practices, and Case Studies* (Palgrave, 2016); *Global Engagement at US Community Colleges,* (2012); and *Community College Models: Globalization and Higher Education Reform* (Spring, 2009).

Dr. Duncan MacLellan is Associate Professor in the Department of Politics and Public Administration and a member of the Yeates School of Graduate Studies, Ryerson University, Ontario, Canada. Since 2012, Duncan has been Director of the graduate (MA) program in Public Policy and Administration. He holds a Ph.D. from the University of Toronto, specializing in educational administration. Duncan's doctoral thesis, *Two Teachers Associations and the Ontario College of Teachers: A Study of Teacher and State Relations*, explored the changing nature of state and teacher relations in Ontario. Duncan's teaching and research interests include educational politics and policy making at the local and provincial levels, state and teacher relations, and local and urban governance issues. Duncan has published in the following periodicals: *Politics and Religion Journal; International Journal of Learning; Academic Leadership; The Journal of the Society for Socialist Studies; Canadian and International Education; Education and Law Journal; The American Review of Canadian Studies. Examples of Duncan's research and writing can also be found in the following books*: *Deputy ministers in Canada's federal and provincial governments: Comparative and jurisdictional perspectives; Academic service-learning across disciplines; Canadian educational leadership; The dark side of leadership: Identifying and overcoming unethical practices in organizations.*

Jabari Mahiri is a Professor of Education and the William and Mary Jane Brinton Family Chair in Urban Teaching. He is Faculty Director of the Multicultural Urban Secondary English MA and Credential Program, Faculty Advisor and Principal Investigator of the Bay Area Writing Project, and a board member of the National Writing Project. He also was a board member of the American Educational Research Association from 2014 to 2017 and board chair of REALM middle and high schools in Berkeley, California from 2011 to 2017. Two of Dr. Mahiri's recent books are *Digital Tools in Urban Schools: Mediating a Remix of Learning* (2011) and *Deconstructing Race: Multicultural Education Beyond the Color-Bind* (2017). He also is editor of *The First Year of Teaching: Classroom Research to Improve Student Learning* (2014) with Sarah Freedman, and guest editor in 2017 of two special issues of the *Multicultural Education Review* on the theme "Cyberlives: Digital Media and Multicultural Education."

Avi Mendelson is a PhD Candidate in the English Department at Brandeis University. He is currently writing a dissertation, *Shape-Shifting Shakespearean Madnesses*, about some airborne madnesses both in Shakespeare's plays and early modern English culture at large. Two generous Mellon grants helped fund nine months in London and three months in Chicago, during which Avi read medical recipe books about epilepsy, rabies, and vertigo at The Welcome History of Medicine Library, Elizabethan psychiatric records at the Bethlem Royal Psychiatric Hospital, and herbals at The Newberry. Avi also works with Mind UK and Core Arts nonprofit organizations that help those suffering from mental illness. Working at the intersection of early modern studies and Disability studies, Avi hopes to use Shakespeare to foster conversations both about the marginalization of the mentally ill and about contemporary debates in mental illness activism.

Dr. Anthony H. Normore is professor of educational leadership, and department chair of Graduate Education at California State University Dominguez Hills, located in Los Angeles. He holds a Ph.D. from the University of Toronto. Dr. Normore's two-pronged research focusses on (a) the *(mis)*-interpretation and *(mis)*-use of leadership and management in higher education, and (b) leadership development of urban school leaders in the context of ethics and social justice. He is the author of 20+ books including *Foundations of educational leadership: The key to developing excellent and equitable schools (2017,* Routledge, with J.S. Brooks); *Restorative practice meets social justice: Un-silencing the voices of at-promise students* (2017, IAP), and; *Handbook of research on communication, leadership, and conflict resolution* (2016, IGI Global publishers). He has published 200+ book chapters, research reviews, and peer-reviewed articles in professional leadership journals, and 300+ regional, national and international professional conference presentations, symposia, and invited keynote addresses. Dr. Normore serves on various editorial review board. He is the recipient of the AERA 2013 *Bridge People Award* for Leadership for Social Justice SIG, and recipient of 2015 *Willower Award of Excellence in Research* awarded by UCEA Consortium of the Study of Leadership and Ethics in Education.

Jessica Parmar is a moderate-to- severe special education teacher at Redondo Union High School, in Redondo Beach, CA. She is currently in my 2nd year of teaching, and was inspired to teach because of how influential educators have been for her growing up. She has personal family connections to disability and mental health which inspired her to teach special education. She teaches within a program called "Essential Skills" at the high school level, which incorporates all core subject areas with an emphasis on functional academics, life skills, independence, and community-based- instruction. Jessica was a credential candidate in special education at California State University Dominguez Hills and worked as Dr. Saili Kulkarni's research assistant, co-publishing a review of research on Augmentative and Alternative Communication. She eventually hopes to attend graduate school and obtain a doctoral degree.

Dr. Gary Rhodes, Ph.D., is the Associate Dean of International Education and Senior International Officer in the College of Extended & International Education at California State University at Dominguez Hills (CSUDH). As Associate Dean & Senior International Officer, he is responsible for providing support for all international initiatives on the CSUDH campus. He is also Director of the Center for Global Education (the Center), which serves as national and international resource to support internationalization of higher education. The Center was founded in 1998 at the University of Southern California, moved to Loyola Marymount University, then UCLA, and now is based at CSUDH. The Center has received various grants and sponsorships over the past 20 years to develop resources for faculty, staff, and students in the U.S. and around the world. Dr. Rhodes areas of expertise include: Internationalization of Higher Education; Study Abroad Program Development and Administration; Integrated International Learning; Diversity and Study Abroad; and, Health, Safety, Crisis and Risk Management for Study Abroad. He publishes articles and presents widely at conferences across the U.S. and around the world. He holds a BA in English from the University of California at Santa Barbara, an MA in International Relations, MSEd. In Education and a PhD in Education, from the University of Southern California. He served in the field as a Technical Advisor in support of a primary education project in Cameroon and served as a Fulbright Specialist in India and South Africa. Dr. Rhodes can be contacted at: *grhodes@csudh.edu*

Dr. Heather M. Rintoul is Associate Professor in the Schulich School of Education, Nipissing University, Ontario, Canada. Dr. Rintoul holds a Ph.D. from the University of Toronto, specializing in *Educational Administration.* Her graduate teaching (doctoral and masters), writing and research focuses on values, ethics and leadership, reflective practice, organizational management, and ethical decision-making. She is the Executive Director (International) of the Consortium for the Study of leadership and Ethics in Education (CSLEE), Director of the *Nipissing University Center for the Study of Leadership and Ethics* (NUCSLE), one of 10 centers worldwide of CSLEE part of the umbrella organization UCEA International, and the Editor-in-Chief of the *Journal of Authentic Leadership in Education.* She has published in periodicals: *Educational Management, Administration & Leadership* (EMAL); *Values, Ethics, and Educational Administration* (VEEA); *Journal of Authentic Leadership in Education* (JALE). Examples of her writing and research can also be found in the following books: *Global perspectives on educational leadership reform*, Emerald; *Examining the assistant principalship*; *Pathways to excellence*; *Handbook of research on effective communication, leadership, and conflict resolution; The dark side of leadership: Identifying and overcoming unethical practices in organizations.* Dr. Rintoul is the recipient of the 2017 Authentic Leadership award from the Consortium of the Study of Leadership and Ethics in Education (CSLEE).

Dr. Albert D. Ritzhaupt is an Associate Professor of Educational Technology in the School of Teaching and Learning at the University of Florida. Dr. Ritzhaupt is an accomplished educational researcher and technologist. Dr. Ritzhaupt has published more than 80 journal articles, book chapters, and conference proceedings; and has presented more than 100 presentations at state, national, and international conferences. His primary research areas focus on the design and development of technology-enhanced learning environments, and technology integration in education, particularly focusing on the factors that facilitate and hinder technology use (e.g., Digital Divide) in formal educational settings.

Dr. Hamoud Salhi is Interim Dean of Undergraduate Studies and Professor of Political Science. His research is in international politics and comparative politics with the Middle East and Northern Africa as his area of study. Dr. Salhi's research focuses on the impact of information technology on society, politics, security and economy in the Middle East and Northern Africa. His publications include a book chapter on *Assessing the Theory of Information Technology and Security for the Middle East* in Johan Eriksson and Giampiero Giacomello's International Relations and Security in the Digital Age; an article, *The State Still Governs,* in International Studies Review; a book review in Review Policy Research journal of Milton Miller's book Networks and States: The Global Politics of Internet Governance.

Dr. Ann Selmi has taught for 45 years, and most of her career has have been spent working with young children with or without disabilities and their teachers. For the past twenty-three years she has taught teachers for the California State University system. Her research focuses on the relationship between the development of language and play. She also worked for five years as a researcher on the cochlear implant program for young deaf children at the House Ear Institute in Los Angeles, and she spent one year as a visiting researcher at the National Institute of Child Health and Development in the Child and Family Research Section in Bethesda, Maryland. Along with a number of published articles, she is the primary author of *Early Childhood Curriculum for All Learners: Integrating Play and Literacy Activities* (2014, Sage Publishers). In 2016, she was awarded a US Department of Education multi-million dollar grant for 5 years to prepare early childhood special education teachers to address some of the most challenging situations in the Los Angeles area.

Dr. Mark Warschauer is Professor of Education and Informatics at the University of California, Irvine and director of both the Teaching & Learning Research Center and the Digital Learning Lab at the University. Previously the founding editor of the *Language Learning & Technology* journal, Dr. Warschauer is now the inaugural editor of *AERA Open.* His research focuses on the use of information and communication technologies for language and literacy development among diverse learners. He is the author or editor of 12 books and more than 100 peer-reviewed papers on this topic, including *Laptops and Literacy* (Teachers College Press); *Technology and Social Inclusion* (MIT Press); and the co-edited volume, *Network-Based Language Teaching* (Cambridge University Press).

Dr. Steven C. Williams received a PhD for history from UCLA in 1990s. Subsequently, he became interested in online learning during his years as an adjunct and lecturer where he became an adherent to online teaching practices even for his face-to-face courses. He learned quickly that online traveled well. Eventually his early-adopter approach to teaching and learning technologies garnered the attention of Warren Ashley—a pioneer and giant of the television educational era who had created an engaging approach to educational television called DHTV that continues to this day to broadcast throughout Southern California to more than 3 million homes. It was Ashley who recruited him to develop a browser-based internet learning management support structure for television courses which he broadcast from studios at California State University Dominguez Hills. In that capacity, he helped administer and guide the integration of the LMS into the fabric of teaching and learning at CSUDH. He currently supports educators who wish to enhance and extend their own teaching using learning management systems and other tools. He is a frequent traveler to Brazil which is the focus of his doctoral research. He can be reached at stevewil@gmail.com.

Dr. Meng Zhao is a marketing professor at California State University, Dominguez Hills. His research interests are in the areas of marketing strategies, innovation management, complex systems, and supply chain management. Currently, he is working on issues such as new product launch strategies, network heterogeneity in new product diffusion processes, wealth effects of NPD announcements, management of organization knowledge, etc. He has published at *Academy of Management Journal, IEEE Transactions on Engineering Management, Journal of Product Innovation Management,* and *Journal of Business Logistics,* among others.

Doron Zinger is a former high school science teacher and school administrator. He has taught at the undergraduate and graduate levels at the University of California, Irvine and California State University Dominguez hills. He is currently a doctoral candidate at UC Irvine. Mr. Zinger has taught classes on the use of technology in instruction. Mr. Zinger has published both journal articles and chapters in edited books in the areas of teacher learning and teacher use of technology to promote student learning. His research interests include teacher preparation, science teacher learning and teaching, and teaching through and with technology, especially in high-need schools.

Dr. Kendall Zoller is an author, educator, researcher, international presenter, and co-author of *The Choreography of Presenting: The 7 Essential Abilities of Effective Presenters* (2010, Corwin Press), and president of Sierra Training Associates specializing in communicative intelligence and Hacking Leadership. He has authored over three dozen reviewed book chapters and journal articles spanning topics of communication, community, and leadership for educators and law enforcement. His work on leadership and presentation skills takes him to schools, districts, universities, state agencies, and corporations across the United States, and Canada and into Europe, China, Thailand, Malaysia, the Philippines and other points east. His lectures, presentations, and paper presentations include the campuses of Harvard, UC Berkeley, St. Anselm College, Boston University, University of Chicago, and Loyola University Maryland. Kendall has a doctorate in Educational Leadership a Masters in Educational Management. Kendall can be reached at kvzollerci@gmail.com

Printed in the United States
By Bookmasters